Microhydrodynamics and Complex Fluids

Dominique Barthès-Biesel

CRC Press
Taylor & Francis Group
Boca Raton London New York

CRC Press is an imprint of the
Taylor & Francis Group, an **informa** business

CRC Press
Taylor & Francis Group
6000 Broken Sound Parkway NW, Suite 300
Boca Raton, FL 33487-2742

First issued in paperback 2017

© 2012 by Taylor & Francis Group, LLC
CRC Press is an imprint of Taylor & Francis Group, an Informa business

No claim to original U.S. Government works

Version Date: 20120424

ISBN 13: 978-1-138-07240-4 (pbk)
ISBN 13: 978-1-4398-8196-5 (hbk)

Visit the Taylor & Francis Web site at
http://www.taylorandfrancis.com

and the CRC Press Web site at
http://www.crcpress.com

Contents

viii

List of Figures

List of Tables

Foreword

The book deals with low Reynolds number flows in which inertia effects are negligible compared to viscous effects. The flow then results from a balance between viscous and pressure forces, and can be described by the Stokes equations, a simple linear form of the Navier–Stokes equations. Stokes flows are common in various industrial, biophysical or natural processes, particularly when the flow scale is small, as in

- The thin lubrication film between two solid elements in relative motion (joints between bones, fluid bearings);
- Narrow pores (porous media, microfluidic networks, blood micro-circulation);
- Thin liquid films (teardrop film, paint or varnish layer);
- Flow around small particles (mud, emulsion, composite material processing);
- Micro-organism swimming.

However, Stokes flows also occur on a large scale when the fluid viscosity is high or when the velocity is low such as in

- Geophysical flows (glacier advance, lava spread, earth plasma motion);
- Gravity currents, spreading of pollution.

The objective of this book is to present the specific phenomena occurring in small Reynolds number flows and to illustrate their consequences in different cases. Each situation is illustrated with examples that can be solved analytically so that the main physical phenomena appear clearly. The modern techniques used for numerical modelling are also presented.

As the viscous behaviour of the fluid plays an essential role, two chapters deal with the flow of 'complex fluids', presented first with the formal analysis developed for the mechanics of suspensions and then with the phenomenological tools of non-Newtonian fluid mechanics.

Each chapter is illustrated with case studies that tackle a variety of very different situations which are a testimony to the wide spectrum of applications of low Reynolds number flows. The case studies are presented in the form of homework or exam problems designed to be solved within one to two hours.

The book is intended for university students who specialise in chemical or mechanical engineering, material science, bioengineering or physics of condensed matter. Elementary notions on continuum mechanics are useful (stress and rate of deformation tensors). However, this book is self-contained and the necessary notions are presented in the first chapter.

Acknowledgements

The author would like to thank all her Ecole Polytechnique colleagues who have contributed to the composition of interesting and feasible exam problems on various topics. Special thanks are in order for Johann Walter and Jean-Paul Barthès, who took on the pain of tracking the bugs and typos in the text.

About the Author

Dominique Barthès-Biesel graduated from Ecole Centrale Paris and then earned a PhD. degree in chemical engineering from Stanford University. She has been a professor at both Ecole Polytechnique and at Compiègne University of Technology, where she taught various classes in classical and complex fluid mechanics, biomechanics, and microfluidics.

Prof. Barthès-Biesel's field of interest is fluid mechanics with a special emphasis on suspensions of deformable particles such as drops, cells, and capsules. She is well-known for her pioneering work on the motion and deformation of encapsulated droplets. She has directed 27 PhD. theses, published over 70 papers, and also worked on industrial projects.

Symbol Description

The main symbols which are used throughout the text are listed below.

c_p	Compactness of a porous medium	S_∞	Surface far from the origin
c	Volume concentration	t	Time
$\mathcal{D}$	Fluid domain	$\mathbf{T}^{ext}$	Traction per unit area exerted on the fluid boundary
$\partial\mathcal{D}$	Boundary of fluid domain		
$\mathbf{e}, e_{ij}$	Shear rate (or rate of deformation) tensor	$\mathbf{u}, u_i$	Velocity vector
		$\mathbf{U}, U_i$	Velocity on boundary
$\mathbf{f}, f_i$	Body force per unit volume	U, V, W	Characteristic velocity scale
g	Gravity acceleration	$\mathbf{x}, x_i$	Position of a point
$\mathbf{I}$	Identity matrix	δ_{ij}	Kronecker symbol
K	Kinetic energy	ε_{ijk}	Permutation symbol
k_p	Permability	ε	Aspect ratio of a film
L	Characteristic length scale	γ_s	Surface tension
N_1, N_2	Normal stress difference	$\dot{\gamma}$	Shear rate
p	Pressure including hydrostatic forces	$\boldsymbol{\omega}, \omega_i$	Vorticity vector
		$\boldsymbol{\Omega}, \Omega_i$	Solid body rotation vector
p'	Hydrodynamic pressure	Φ	Rate of energy dissipation
$\mathcal{P}_V^{ext}$	Rate of work of external volume forces	Φ_L	Potential function
		ϕ_p	Porosity
$\mathcal{P}_S^{ext}$	Rate of work of external surface forces	Ψ	Stream function
		μ	Dynamic viscosity
Pe	Péclet number	ν	Kinematic viscosity
Q	Flow rate	ρ	Density
r	Distance from origin	$\boldsymbol{\sigma}, \sigma_{ij}$	Stress tensor including hydrostatic forces
$\Re$	Real part		
Re	Reynolds number	$\boldsymbol{\sigma}', \sigma'_{ij}$	Hydrodynamic stress tensor
St	Stokes number	$\boldsymbol{\tau}, \tau_{ij}$	Deviatoric stress tensor
		τ	Shear stress

1

Fundamental Principles

CONTENTS

We present in this chapter the fundamental concepts that will be useful for understanding the rest of the book. The different equations (mass conservation, equation of motion, etc) will be simply stated. The demonstrations can be found in any classical textbook on fluid mechanics.

We also introduce the main notations. The fluid will be considered as an incompressible continuous medium with constant and uniform density ρ. When the fluid is set into motion, the velocity field $\mathbf{u}(\mathbf{x}, t)$ and the the pressure field $p(\mathbf{x}, t)$ are continuous functions of position $\mathbf{x}$ and time t. This hypothesis implies that the scale of study of the fluid will be much larger than the scale of inhomogeneities such as particles in suspensions or macromolecules in polymer solutions. We normally use a Galilean reference frame with a Cartesian coordinate system.

Finally, we often use the index notation with implicit summation on repeated indices (the so-called Einstein convention presented in detail in Appendix A.1). To facilitate reading, the equations are often written both with the vector and with the index notations.

1.1 Mass Conservation

Mass conservation implies a local relation between ρ and $\mathbf{u}$

$$\frac{\partial \rho}{\partial t} + \nabla \cdot (\rho \mathbf{u}) = 0 \qquad \text{or} \qquad \frac{\partial \rho}{\partial t} + \frac{\partial (\rho u_i)}{\partial x_i} = 0 \qquad (1.1)$$

For an incompressible fluid, the density is constant. Mass conservation then becomes the so-called continuity equation

$$\nabla \cdot \mathbf{u} = \frac{\partial u_i}{\partial x_i} = 0 \tag{1.2}$$

1.2   Equation of Motion

Newton's law of mechanics applied to a continuous medium leads to the equation of motion

$$\rho \frac{D\mathbf{u}}{Dt} = \mathbf{f} + \nabla \cdot \boldsymbol{\sigma}' \qquad \text{or} \qquad \rho \frac{Du_i}{Dt} = f_i + \frac{\partial \sigma'_{ij}}{\partial x_j} \tag{1.3}$$

where the convective derivative D/Dt is defined by

$$\frac{D}{Dt} = \frac{\partial}{\partial t} + \mathbf{u} \cdot \nabla \qquad \text{or} \qquad \frac{D}{Dt} = \frac{\partial}{\partial t} + u_i \frac{\partial}{\partial x_i} \tag{1.4}$$

The Cauchy stress tensor $\boldsymbol{\sigma}'(\mathbf{x}, t)$ represents the internal force per unit area acting in the fluid at point $\mathbf{x}$ at time t. The external force per unit volume acting on the fluid is noted $\mathbf{f}(\mathbf{x}, t)$. When the external force is conservative, it is related to a potential function V by

$$\mathbf{f} = -\nabla V \qquad \text{or} \qquad f_i = -\frac{\partial V}{\partial x_i} \tag{1.5}$$

We can then define a modified stress tensor

$$\boldsymbol{\sigma} = \boldsymbol{\sigma}' - V\mathbf{I} \qquad \text{or} \quad \sigma_{ij} = \sigma'_{ij} - V\delta_{ij} \tag{1.6}$$

The usual case concerns gravity forces for which

$$\mathbf{f} = \rho \mathbf{g} \tag{1.7}$$

where $\mathbf{g}$ is the acceleration of gravity. In this case, the modified stress tensor includes hydrostatic effects

$$\boldsymbol{\sigma} = \boldsymbol{\sigma}' + \rho \mathbf{g} \cdot \mathbf{x} \, \mathbf{I} \qquad \text{or} \qquad \sigma_{ij} = \sigma'_{ij} + \rho \, g_k x_k \, \delta_{ij} \tag{1.8}$$

and the equation of motion of the fluid becomes simply

$$\rho \frac{D\mathbf{u}}{Dt} = \nabla \cdot \boldsymbol{\sigma} \qquad \text{or} \qquad \rho \frac{Du_i}{Dt} = \frac{\partial \sigma_{ij}}{\partial x_j} \tag{1.9}$$

1.3   Newtonian Fluid

The rate of strain (also called rate of deformation or shear rate) tensor **e** is defined by

$$\mathbf{e} = \frac{1}{2} \left[\nabla \mathbf{u} + {}^{\mathsf{T}}(\nabla \mathbf{u}) \right] \qquad \text{or} \qquad e_{ij} = \frac{1}{2} \left(\frac{\partial u_i}{\partial x_j} + \frac{\partial u_j}{\partial x_i} \right) \qquad (1.10)$$

where the superscript ^{T} denotes the transpose of a matrix. For an incompressible fluid, the continuity Equation (1.2) leads to

$$tr(\mathbf{e}) = e_{11} + e_{22} + e_{33} = 0 \qquad \text{or} \qquad e_{kk} = 0 \qquad (1.11)$$

The fluid constitutive law relates the stress tensor to the rate of strain tensor in the fluid. An incompressible Newtonian fluid is defined by a linear relation between the Cauchy stress and the rate of strain tensor

$$\boldsymbol{\sigma}' = -p'\mathbf{I} + 2\mu\mathbf{e} \qquad \text{or} \qquad \sigma'_{ij} = -p'\delta_{ij} + 2\mu e_{ij} \qquad (1.12)$$

where $p'(\mathbf{x}, t)$ is the hydrodynamic pressure in the fluid. The pressure is an unknown scalar field that appears because of the incompressibility constraint. It must be determined as part of the solution of any fluid motion problem. The fluid viscosity μ (Pa s) is assumed to be constant. For example, the viscosity of water is 1.005×10^{-3} Pa s at 20°C. We can also use the kinematic viscosity ν (m²/s)

$$\nu = \mu/\rho \qquad (1.13)$$

The kinematic viscosity of water is 1.004×10^{-6} m²/s at 20°C.

When the volume forces are due to gravity only, we can introduce the modified pressure

$$p = p' - \rho g_i x_i + Cst \qquad (1.14)$$

which corresponds to the sum of the hydrodynamic pressure p' and of the hydrostatic pressure. Newton's constitutive law can then be written as

$$\boldsymbol{\sigma} = -p\mathbf{I} + \boldsymbol{\tau} \qquad \text{with} \qquad \boldsymbol{\tau} = 2\mu\mathbf{e} \qquad (1.15)$$

where $\boldsymbol{\tau}$ is a tensor with zero trace (tr $\boldsymbol{\tau} = 0$) which represents the contribution of viscous effects to the stress tensor.

Remark

There exist in nature only two kinds of fluids: *Newtonian fluids*, which satisfy Newton constitutive law, and *non-Newtonian fluids*, which have a more complex constitutive law (Chapter 11). Schematically, we can consider that pure fluids (e.g. air, water, low density oil) have a Newtonian behaviour when subjected to moderate shear rates. *Complex fluids* (e.g. suspensions of deformable or rigid particles, polymer solutions, blood) do not satisfy Newton's constitutive law for a given range of shear rates. The study of such fluids is called *rheology* and is presented in more detail in Chapter 11.

1.4   Navier–Stokes Equations

Combining Equations (1.2), (1.9) and (1.15), we obtain the equation of motion for an incompressible Newtonian fluid

$$\rho\frac{D\mathbf{u}}{Dt} = \rho\left(\frac{\partial\mathbf{u}}{\partial t} + (\mathbf{u}\cdot\nabla)\mathbf{u}\right) = -\nabla p + \mu\nabla^2\mathbf{u} \qquad (1.16)$$

or

$$\rho\frac{Du_i}{Dt} = \rho\left(\frac{\partial u_i}{\partial t} + u_k\frac{\partial u_i}{\partial x_k}\right) = -\frac{\partial p}{\partial x_i} + \mu\frac{\partial^2 u_i}{\partial x_k\partial x_k} \qquad (1.17)$$

The term $\partial\mathbf{u}/\partial t$ represents the acceleration of a fluid particle due to time variations of velocity (e.g. pulsatile flow, sudden acceleration). This term is zero for a steady flow (that is, a constant flow that does not depend on time). The term $\mathbf{u}\cdot\nabla\mathbf{u}$ represents the convective acceleration of a particle due to spatial variations of velocity. For example, in the flow through a convergent-divergent channel, the fluid particles accelerate in the convergent part and decelerate in the divergent part even if the flow is steady. The term ∇p represents the pressure gradient due to the fluid incompressibility. Finally, the term $\mu\nabla^2\mathbf{u}$ represents the viscous forces per unit volume which act inside the fluid and slow it down.

Of course, one should not forget to associate the mass conservation equation (1.2)

$$\nabla\cdot\mathbf{u} = \frac{\partial u_i}{\partial x_i} = 0$$

The two Equations (1.16) and (1.2) are the well-known Navier–Stokes equations which model the flow of any Newtonian incompressible fluid. These nonlinear equations are difficult to solve in general, but give a good description of many engineering or natural flows.

1.5   Energy Dissipation

We now perform an energy balance on the fluid. We thus consider a fluid flowing with velocity $\mathbf{u}(\mathbf{x}, t)$ in a fixed domain $\mathcal{D}$ with boundary $\partial\mathcal{D}$ on which the velocity is $\mathbf{U}(\mathbf{x}, t)$ (Figure 1.1)

$$\mathbf{u}(\mathbf{x}) = \mathbf{U}(\mathbf{x}) \quad \text{or} \quad u_i(\mathbf{x}) = U_i(\mathbf{x}) \quad \text{for} \quad \mathbf{x}\in\partial\mathcal{D} \qquad (1.18)$$

The total kinetic energy of the flow is

$$K = \frac{1}{2}\rho\int_{\mathcal{D}} u_i u_i\, dV \qquad (1.19)$$

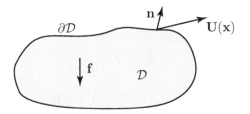

FIGURE 1.1
Definition of a fluid domain $\mathcal{D}$ bounded by $\partial\mathcal{D}$.

The rate of variation of the total kinetic energy is then

$$\frac{dK}{dt} = \rho \int_{\mathcal{D}} u_i \frac{Du_i}{Dt} dV \tag{1.20}$$

The equation of motion (1.3) relates the acceleration $D\mathbf{u}/Dt$ to the stress field in the fluid

$$\frac{dK}{dt} = \int_{\mathcal{D}} u_i \left[f_i + \frac{\partial \sigma'_{ij}}{\partial x_j} \right] dV = \int_{\mathcal{D}} u_i f_i dV + \int_{\mathcal{D}} u_i \frac{\partial \sigma'_{ij}}{\partial x_j} dV \tag{1.21}$$

We use the chain derivative rule to evaluate the last term on the right-hand side

$$\int_{\mathcal{D}} u_i \frac{\partial \sigma'_{ij}}{\partial x_j} dV = \int_{\mathcal{D}} \frac{\partial (u_i \sigma'_{ij})}{\partial x_j} dV - \int_{\mathcal{D}} \sigma'_{ij} \frac{\partial u_i}{\partial x_j} dV$$
$$= \int_{\partial\mathcal{D}} u_i \sigma'_{ij} n_j \, dS - \int_{\mathcal{D}} \sigma'_{ij} \, e_{ij} dV \tag{1.22}$$

where $\mathbf{n}$ is the unit normal vector to the domain boundary, pointing out of $\mathcal{D}$. In Equation (1.22), we have used the fact that σ'_{ij} is symmetric to replace the product $\sigma'_{ij} \, \partial u_i/\partial x_j$ by $\sigma'_{ij} \, e_{ij}$ and we have used Gauss' theorem to convert the volume integral into a surface integral. We finally obtain

$$\frac{dK}{dt} = \int_{\mathcal{D}} u_i f_i dV + \int_{\partial\mathcal{D}} U_i T_i^{ext} dS - \int_{\mathcal{D}} \sigma'_{ij} \, e_{ij} dV \tag{1.23}$$

where $T_i^{ext} = \sigma'_{ij} n_j$ is the traction per unit area exerted *on* the fluid at the boundary $\partial\mathcal{D}$. The final energy balance thus becomes

$$\frac{dK}{dt} = \mathcal{P}_V^{ext} + \mathcal{P}_S^{ext} - \Phi \tag{1.24}$$

where $\mathcal{P}_V^{ext} = \int_{\mathcal{D}} u_i f_i dV$ and $\mathcal{P}_S^{ext} = \int_{\partial\mathcal{D}} U_i T_i^{ext} dS$ represent the rate of work of the volume and surface external forces, respectively. The rate of energy dissipation Φ is defined as

$$\Phi = \int_{\mathcal{D}} \sigma'_{ij} \, e_{ij} dV \tag{1.25}$$

When the fluid is Newtonian, we can replace σ'_{ij} by the expression given by Newton's law (1.12) and obtain

$$\Phi = 2\mu \int_{\mathcal{D}} e_{ij}\, e_{ij}\, dV \tag{1.26}$$

where we have used the relation $e_{ij}\delta_{ij} = 0$. It follows that the energy dissipation Φ is always positive or zero. Equation (1.24) allows us to determine the loss of kinetic energy due to viscous friction for any fluid.

1.6 Dimensional Analysis

We use the following scales for the flow:

- L for length.
- U for velocity.
- T for the characteristic time of variation of the flow.

A dimensional analysis of the different terms in the Navier–Stokes equations allows us to determine the following orders of magnitude

$$\left|\rho\frac{\partial \mathbf{u}}{\partial t}\right| \sim \frac{\rho U}{T}, \qquad |\rho\mathbf{u}.\nabla\mathbf{u}| \sim \frac{\rho U^2}{L}, \qquad |\mu\nabla^2\mathbf{u}| \sim \frac{\mu U}{L^2} \tag{1.27}$$

where the notation $\sim$ means here 'order of magnitude'. The pressure gradient (always present in an incompressible fluid) balances the largest term in Equation (1.16). To compare the orders of magnitude of the different terms, we take their ratio

$$\frac{|\rho\mathbf{u}.\nabla\mathbf{u}|}{|\mu\nabla^2\mathbf{u}|} \sim \frac{\rho U L}{\mu}, \qquad \frac{|\rho\partial\mathbf{u}/\partial t|}{|\mu\nabla^2\mathbf{u}|} \sim \frac{\rho L^2}{T\mu} \tag{1.28}$$

and thus obtain the Reynolds number Re

$$Re = \frac{\rho U L}{\mu} = \frac{U L}{\nu} \tag{1.29}$$

which measures the ratio between inertial and viscous forces. The Stokes number St is defined by

$$St = \frac{\rho L^2}{\mu T} = \frac{L^2}{\nu T} \tag{1.30}$$

and measures the ratio between the transient inertia effects and the viscous effects. When μ is large or when L is small, St is small and viscous effects are dominant. For a stationary flow, T is infinite and $St = 0$.

2

General Properties of Stokes Flows

CONTENTS

When the Reynolds and Stokes numbers are small ($Re \ll 1$ and $St \ll 1$), the inertia terms can be neglected in the Navier–Stokes equations. We then obtain the Stokes equations which are linear. This linearity has some important consequences regarding the dynamics of the flow as shown in this chapter.

2.1   Stationary Stokes Equations

Stokes flows occur when the Reynolds number is very small, that is,

$$Re \ll 1 \tag{2.1}$$

This means that the viscous forces are much larger than the convective inertia forces. The convective acceleration term $\rho \mathbf{u} \cdot \nabla \mathbf{u}$ can thus be neglected in the Navier–Stokes equations. Furthermore, when the Stokes number is very small, $St \ll 1$, or when the flow is steady, the full acceleration term $\rho D\mathbf{u}/Dt$ is negligible and the Navier–Stokes momentum equation becomes the Stokes momentum equation

$$\nabla \cdot \boldsymbol{\sigma} = 0 \qquad \text{or} \qquad \frac{\partial \sigma_{ik}}{\partial x_k} = 0 \tag{2.2}$$

In this book, we consider only steady Stokes flows described by Equation (2.2).

2.1.1   Pressure–Velocity Relation

Replacing $\boldsymbol{\sigma}$ by its value for a Newtonian fluid

$$\boldsymbol{\sigma} = -p\mathbf{I} + 2\mu\mathbf{e} \qquad \text{or} \qquad \sigma_{ij} = -p\delta_{ij} + 2\mu e_{ij} \tag{2.3}$$

we obtain a first relation between the velocity and the pressure

$$0 = -\nabla p + \mu \nabla^2 \mathbf{u} \quad \text{or} \quad \frac{\partial p}{\partial x_i} = \mu \frac{\partial^2 u_i}{\partial x_k \partial x_k} \tag{2.4}$$

to which we must add the continuity equation (always!)

$$\nabla \cdot \mathbf{u} = 0 \tag{2.5}$$

The physical meaning of Equation (2.4) is that the pressure gradient in the fluid is balanced by the viscous friction at any time t. Equations (2.4) and (2.5) constitute the Stokes equations for a Newtonian fluid.

Taking the divergence of Equation (2.4) and accounting for the incompressibility condition (2.5), we obtain

$$\nabla^2 p = 0 \tag{2.6}$$

It follows that the pressure is a harmonic function which satisfies the Laplace equation. We can express the velocity field in the fluid as

$$\mathbf{u} = \frac{p}{2\mu} \mathbf{x} + \mathbf{u}^{(H)} \tag{2.7}$$

Equation (2.4) imposes that $\mathbf{u}^{(H)}$ should also be a harmonic function which satisfies

$$\nabla^2 \mathbf{u}^{(H)} = 0 \tag{2.8}$$

The continuity equation leads to another condition for $\mathbf{u}^{(H)}$

$$\nabla \cdot \mathbf{u}^{(H)} = -\frac{1}{2\mu}\,(3p + \mathbf{x} \cdot \nabla p) \tag{2.9}$$

We can thus express $\mathbf{u}$ and p as sums of fundamental solutions of the Laplace equation as shown in Section 2.7.

2.1.2 Pressure–Vorticity Relation

The vorticity $\boldsymbol{\omega}$ is defined as the curl of velocity

$$\boldsymbol{\omega} = \nabla \times \mathbf{u} \quad \text{or} \quad \omega_i = \varepsilon_{ijk}\frac{\partial u_k}{\partial x_j} \tag{2.10}$$

where ε_{ijk} is the permutation symbol (Appendix A.1). Taking the curl of Equation (2.4), we obtain the equation of motion for $\boldsymbol{\omega}$

$$\nabla^2 \boldsymbol{\omega} = 0 \quad \text{or} \quad \frac{\partial^2 \omega_i}{\partial x_k \partial x_k} = 0 \tag{2.11}$$

We can thus relate the pressure to the vorticity instead of the velocity. In order to do this, we recall the classical vector identity

$$\nabla^2 \mathbf{u} = -\nabla \times (\nabla \times \mathbf{u}) + \nabla \cdot (\nabla \cdot \mathbf{u}) \tag{2.12}$$

In view of Equation (2.5), this simplifies to

$$\nabla^2 \mathbf{u} = -\nabla \times (\nabla \times \mathbf{u}) \tag{2.13}$$

Using Equation (2.13), we can write the Stokes momentum Equation (2.4) under the alternate form

$$\nabla p = -\mu \nabla \times \boldsymbol{\omega} \quad \text{or} \quad \frac{\partial p}{\partial x_i} = -\mu\,\varepsilon_{ijk}\frac{\partial \omega_k}{\partial x_j} \tag{2.14}$$

to which we must not forget to add the mass conservation Equation (2.5)!

2.1.3 Boundary Conditions on a Solid Surface

The fluid particles that are in contact with a solid wall remain there and move with it. As a consequence, the fluid velocity is equal to the wall velocity. Let $F(\mathbf{x}, t) = 0$ be the equation of the wall and $\mathbf{U}(\mathbf{x}, t)$ its velocity. The no-slip condition is then

$$\mathbf{u}(\mathbf{x}, t) = \mathbf{U}(\mathbf{x}, t) \quad \text{for } \mathbf{x} \text{ such that } F(\mathbf{x}, t) = 0 \tag{2.15}$$

Another consequence is that the wall is a material surface which is convected by the motion

$$\frac{DF}{Dt} = 0 \qquad (2.16)$$

or

$$\frac{\partial F}{\partial t} + \mathbf{U} \cdot \nabla F = 0 \qquad (2.17)$$

The term ∇F is directed along the unit normal vector $\mathbf{n}$ to the surface and can thus be written as $\nabla F = \mathbf{n}|\nabla F|$. Using Equation (2.15), we obtain the kinematic condition on a moving surface

$$\frac{1}{|\nabla F|}\frac{\partial F}{\partial t} + \mathbf{U} \cdot \mathbf{n} = 0 \qquad (2.18)$$

which relates the time evolution of the surface shape to its velocity.

2.2   Simple Stokes Flow Problem

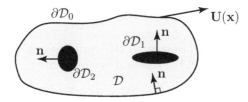

FIGURE 2.1
Stokes flow in the fluid domain $\mathcal{D}$ bounded by $\partial\mathcal{D}_0 \cup \partial\mathcal{D}_1 \cup \partial\mathcal{D}_2$.

A typical Stokes flow problem corresponds to the flow of a Newtonian incompressible fluid with constant viscosity μ and density ρ, in a domain $\mathcal{D}$ with exterior boundary $\partial\mathcal{D}_0$ (this exterior boundary may be 'infinitely' far from the centre of $\mathcal{D}$). The domain $\mathcal{D}$ may contain n internal subdomains with respective boundaries $\partial\mathcal{D}_1$, $\partial\mathcal{D}_2$, ... (Figure 2.1). The overall boundary $\partial\mathcal{D}$ of domain $\mathcal{D}$ thus consists of

$$\partial\mathcal{D} = \partial\mathcal{D}_0 \cup \partial\mathcal{D}_1 \cup ... \cup \partial\mathcal{D}_n \qquad (2.19)$$

We assume that we know the velocity $\mathbf{u}$ on each point of $\partial\mathcal{D}$

$$\mathbf{u}(\mathbf{x}, t) = \mathbf{U}(\mathbf{x}, t) \quad \text{for} \quad \mathbf{x} \in \partial\mathcal{D} \qquad (2.20)$$

where $\mathbf{U}(\mathbf{x}, t)$ is a piece-wise continuous function. Mass conservation Equation (2.5) requires

$$\int_{\partial\mathcal{D}} \mathbf{U}(\mathbf{x}, t).\mathbf{n} \, dS = 0 \qquad (2.21)$$

where **n** is the unit vector normal to $\partial \mathcal{D}$, which usually points out of the inclusion and thus points into the fluid.

Although the explicit time dependency $\rho D\mathbf{u}/Dt$ has been eliminated from the equations of motion, this does not mean that a Stokes flow is necessarily steady. Indeed, the flow depends on time through the motion $\mathbf{U}(\mathbf{x}, t)$ of the boundaries. For example, a sphere sedimenting in a quiescent fluid creates an unsteady disturbance flow $\mathbf{u}(\mathbf{x}, t)$ in its vicinity and in its wake. The equation of motion implies that at each time, the pressure forces balance the viscous dissipation forces. The fluid thus goes through a succession of equilibrium states and the flow is said to be *quasi-steady*.

We now seek the flow field that satisfies the Stokes Equations (2.4) and (2.5) with boundary condition (2.20). The solution to this general problem has some remarkable properties which are presented in the next sections.

2.3 Linearity and Reversibility

A major property of the Stokes equations as compared to the Navier–Stokes equations is their linearity. It follows from the elimination of the non-linear convective acceleration term.

The first consequence is that a Stokes problem can be decomposed into a linear combination of subproblems.

2.3.1 Linearity Theorem

Let $\mathbf{u}^{(1)}(\mathbf{x})$, $p^{(1)}(\mathbf{x})$ and $\mathbf{u}^{(2)}(\mathbf{x})$, $p^{(2)}(\mathbf{x})$ be two solutions of Equations (2.4) and (2.5) which satisfy the two boundary conditions, respectively,

$$\mathbf{u}^{(1)} = \mathbf{U}^{(1)}(\mathbf{x}) \quad \text{and} \quad \mathbf{u}^{(2)} = \mathbf{U}^{(2)}(\mathbf{x}) \quad \text{for} \quad \mathbf{x} \in \partial \mathcal{D} \qquad (2.22)$$

The flow field $\mathbf{u} = \lambda_1 \mathbf{u}^{(1)}(\mathbf{x}) + \lambda_2 \mathbf{u}^{(2)}(\mathbf{x})$, $p = \lambda_1 p^{(1)}(\mathbf{x}) + \lambda_2 p^{(2)}(\mathbf{x})$ satisfies Equations (2.4) and (2.5) with the boundary condition

$$\mathbf{u} = \lambda_1 \mathbf{U}^{(1)}(\mathbf{x}) + \lambda_2 \mathbf{U}^{(2)}(\mathbf{x}) \quad \text{for} \quad \mathbf{x} \in \partial \mathcal{D} \qquad (2.23)$$

where λ_1 and λ_2 are two arbitrary scalars. This linearity theorem, being a direct consequence of the equation linearity, has a trivial demonstration.

This remarkable property is used to design numerical solutions of the Stokes equations. The technique consists of finding the linear combination of elementary solutions of Equations (2.4) and (2.5) which allows us to satisfy the boundary conditions of the problem (Chapter 8).

2.3.2   Reversibility

An interesting particular case corresponds to $\lambda_1 = -1$ and $\lambda_2 = 0$. Indeed, when we invert the motion of the boundary ($\mathbf{U}$ becomes $-\mathbf{U}$), the flow field in the fluid is also inverted: $\mathbf{u}(\mathbf{x})$ becomes $-\mathbf{u}(\mathbf{x})$. For example, starting with a fluid in configuration $\mathcal{C}(t)$, if we move the boundary with velocity $\mathbf{U}(\mathbf{x})$ between time t and $t + t'$, the fluid reaches the deformed configuration $\mathcal{C}(t + t')$. Starting from this configuration, if we move the boundary with the opposite velocity $-\mathbf{U}(\mathbf{x})$ for a duration t', then at time $t + 2t'$ we find the same configuration as at time t: $\mathcal{C}(t + 2t') = \mathcal{C}(t)$.

The reversibility of the Stokes equations means that it is difficult to mix two fluids in the Stokes regime, unless we impose a chaotic motion to the system. Another consequence is that it is difficult to swim at low Reynolds number. Indeed, a back-and-forth flapping motion will be totally inefficient because of reversibility. Micro-organisms must thus adapt their propulsion strategy in order to be able to swim (Chapter 6, Section 6.5.2 and Problems 6.6.2 and 6.6.3).

2.4   Uniqueness

Another very important property of a Stokes problem is that the solution is unique. So if we find a solution by whatever means, we know it is the one. This property is now expressed in a more rigorous fashion.

2.4.1   Uniqueness Theorem

Let $\mathbf{u}^{(1)}(\mathbf{x})$, $p^{(1)}(\mathbf{x})$ and $\mathbf{u}^{(2)}(\mathbf{x})$, $p^{(2)}(\mathbf{x})$ be two solutions of Equations (2.4) and (2.5) that both satisfy the same boundary conditions

$$\mathbf{u}^{(1)} = \mathbf{U}(\mathbf{x}) \quad \text{and} \quad \mathbf{u}^{(2)} = \mathbf{U}(\mathbf{x}) \quad \text{for} \quad \mathbf{x} \in \partial \mathcal{D} \tag{2.24}$$

Then

$$\mathbf{u}^{(1)} \equiv \mathbf{u}^{(2)} \quad \text{and} \quad p^{(1)} - p^{(2)} = Cst \quad \text{for} \quad \mathbf{x} \in \mathcal{D} \tag{2.25}$$

2.4.2   Demonstration

We note $e_{ij}^{(\alpha)}$ and $\sigma_{ij}^{(\alpha)}$ ($\alpha = 1, 2$), the shear rate and stress tensors associated to the velocity field $\mathbf{u}^{(\alpha)}$. We then introduce the new velocity field $\Delta \mathbf{u} = \mathbf{u}^{(2)} - \mathbf{u}^{(1)}$ with associated shear rate and stress tensors

$$\Delta e_{ij} = e_{ij}^{(2)} - e_{ij}^{(1)} \quad \text{and} \quad \Delta \sigma_{ij} = \sigma_{ij}^{(2)} - \sigma_{ij}^{(1)} \tag{2.26}$$

In view of the linearity property (Section 2.3), with $\lambda_1 = 1$ and $\lambda_2 = -1$, we find that the field $\Delta \boldsymbol{u}$ also satisfies the Stokes momentum equation

$$\frac{\partial(\Delta \sigma_{ij})}{\partial x_j} = 0 \tag{2.27}$$

with associated boundary condition

$$\Delta \mathbf{u} = 0 \quad \text{for} \quad \mathbf{x} \in \partial \mathcal{D} \tag{2.28}$$

Newton's law allows us to compute $\Delta \sigma_{ij}$:

$$\Delta \sigma_{ij} = -\Delta p \, \delta_{ij} + 2\mu \, \Delta e_{ij} \tag{2.29}$$

We now determine the energy dissipation $\Delta \Phi$ in the domain $\mathcal{D}$:

$$\Delta \Phi = 2\mu \int_{\mathcal{D}} \Delta e_{ij} \Delta e_{ij} \, dV \tag{2.30}$$

Using Equations (2.29) and (1.11), we find

$$\Delta \Phi = \int_{\mathcal{D}} \Delta e_{ij} \Delta \sigma_{ij} \, dV \tag{2.31}$$

or

$$\Delta \Phi = \int_{\mathcal{D}} \frac{\partial(\Delta u_i)}{\partial x_j} \Delta \sigma_{ij} \, dV \tag{2.32}$$

Using the chain derivative rule and Gauss' theorem

$$\Delta \Phi = \int_{\mathcal{D}} \frac{\partial(\Delta u_i)}{\partial x_j} \Delta \sigma_{ij} \, dV = \int_{\partial \mathcal{D}} \Delta u_i \Delta \sigma_{ij} n_j dS - \int_{\mathcal{D}} \frac{\partial(\Delta \sigma_{ij})}{\partial x_j} \Delta u_i dV \tag{2.33}$$

Noting that $\Delta u_i = 0$ on $\partial \mathcal{D}$, we obtain $\Delta \Phi = 0$. Then, using Equation (2.29), we find

$$\Delta \Phi = 0 = 2\mu \int_{\mathcal{D}} \Delta e_{ij} \Delta e_{ij} \, dV \tag{2.34}$$

It follows that $\Delta e_{ij} = 0$ in $\mathcal{D}$ and that

$$e_{ij}^{(1)} = e_{ij}^{(2)} \tag{2.35}$$

which is not yet sufficient to prove that the solution is unique. We thus compute the norm of the vorticity of the field $\Delta \mathbf{u}$ (Appendix A.1)

$$\Delta \omega_k \Delta \omega_k = \varepsilon_{klm} \varepsilon_{kij} \frac{\partial(\Delta u_m)}{\partial x_l} \frac{\partial(\Delta u_j)}{\partial x_i} \tag{2.36}$$

$$= 2 \, \Delta e_{kl} \Delta e_{kl} - 2 \frac{\partial}{\partial x_k} \left[\Delta u_l \frac{\partial(\Delta u_k)}{\partial x_l} \right] \tag{2.37}$$

Integrating in $\mathcal{D}$, we find

$$\int_{\mathcal{D}} \Delta\omega_k \Delta\omega_k \, dV = 2 \int_{\mathcal{D}} \Delta e_{kl} \Delta e_{kl} dV - 2 \int_{\partial\mathcal{D}} \Delta u_l \frac{\partial(\Delta u_k)}{\partial x_l} n_k dS \qquad (2.38)$$

Using Equation (2.34) and $\Delta\mathbf{u} = 0$ for $\mathbf{x} \in \partial\mathcal{D}$, we finally obtain

$$\int_{\mathcal{D}} \Delta\omega_k \Delta\omega_k \, dV = 0 \qquad (2.39)$$

or

$$\omega_i^{(1)} = \omega_i^{(2)} \qquad (2.40)$$

Equations (2.35) and (2.40) thus imply that

$$\mathbf{u}^{(1)} = \mathbf{u}^{(2)} \quad \text{for} \quad \mathbf{x} \in \mathcal{D} \qquad (2.41)$$

Combining Equations (2.4) and (2.41), we find

$$p^{(1)} = p^{(2)} + Cst \quad \text{for} \quad \mathbf{x} \in \mathcal{D} \qquad (2.42)$$

which now allows us to state that the solution to the Stokes equations is indeed unique.

2.5   Minimum Energy Dissipation

For a Stokes flow in which inertia forces are negligible compared to viscous forces, the kinetic energy balance of Equation (1.24) becomes

$$\mathcal{P}_V^{ext} + \mathcal{P}_S^{ext} = \Phi \qquad (2.43)$$

This means that all the energy brought by the external forces is mostly dissipated by the viscous effects and thus increases the fluid temperature. However, an increase in temperature generally leads to a decrease in viscosity. As a consequence, the flow conditions are modified. In some applications (e.g. lubrication, viscometry), it is essential to maintain a constant fluid temperature and thus the calories produced by friction must be removed. This is why a viscometer, which is used to measure the viscosity of a fluid, always includes a system to regulate the temperature.

An important property of Stokes flows is that they correspond to a *minimum energy dissipation*.

2.5.1   Minimum Energy Theorem

Let $\mathbf{u}^{(1)}$ be the unique solution of Equation (2.4) with boundary condition (2.20). Let $\mathbf{u}^{(2)}$ be a 'kinematically compatible' velocity field which satisfies only the continuity Equation (2.5) and the boundary condition (2.20)

$$\nabla \cdot \mathbf{u}^{(2)} = 0 \quad \text{for} \quad \mathbf{x} \in \mathcal{D} \quad \text{and} \quad \mathbf{u}^{(2)} = \mathbf{U}(\mathbf{x}) \quad \text{for} \quad \mathbf{x} \in \partial\mathcal{D} \qquad (2.44)$$

Then the energy dissipations in the two flows are related by

$$\Phi^{(1)} \le \Phi^{(2)} \tag{2.45}$$

where equality occurs only if $\mathbf{u}^{(1)} = \mathbf{u}^{(2)}$.

This result implies that the Stokes flow leads to less energy dissipation than any other flow of an incompressible fluid that satisfies the same boundary conditions in $\mathcal{D}$. In particular, there will be more energy dissipation in a flow that satisfies the Navier–Stokes equations.

2.5.2 Demonstration

We use the same notations as in Section 2.4 and introduce the velocity field $\Delta \mathbf{u} = \mathbf{u}^{(2)} - \mathbf{u}^{(1)}$ which is such that $\Delta \mathbf{u} = 0$ for $\mathbf{x} \in \partial \mathcal{D}$. We compute the energy dissipation difference in the two flows $\mathbf{u}^{(1)}$ and $\mathbf{u}^{(2)}$:

$$\Phi^{(1)} - \Phi^{(2)} = 2\mu \int_{\mathcal{D}} \left[e_{ij}^{(2)} e_{ij}^{(2)} - e_{ij}^{(1)} e_{ij}^{(1)} \right] dV = 2\mu \int_{\mathcal{D}} \Delta e_{ij} \left[e_{ij}^{(1)} + e_{ij}^{(2)} \right] dV$$

$$= 2\mu \int_{\mathcal{D}} \Delta e_{ij} \Delta e_{ij} dV + 4\mu \int_{\mathcal{D}} \Delta e_{ij} e_{ij}^{(1)} dV \tag{2.46}$$

Newton's law allows us to compute the second terms on the right-hand side of Equation (2.46):

$$2\mu \int_{\mathcal{D}} \Delta e_{ij} e_{ij}^{(1)} dV = \int_{\mathcal{D}} \Delta e_{ij} \sigma_{ij}^{(1)} dV = \int_{\mathcal{D}} \frac{\partial (\Delta u_j)}{\partial x_i} \sigma_{ij}^{(1)} dV \tag{2.47}$$

Integrating by parts and using Gauss' theorem, we obtain

$$\int_{\mathcal{D}} \frac{\partial (\Delta u_j)}{\partial x_i} \sigma_{ij}^{(1)} dV = - \int_{\mathcal{D}} \Delta u_j \frac{\partial \sigma_{ij}^{(1)}}{\partial x_i} dV = 0 \tag{2.48}$$

Using Equation (2.47), we find that Equation (2.46) becomes

$$\Phi^{(1)} - \Phi^{(2)} = \int_{\mathcal{D}} \left[e_{ij}^{(2)} e_{ij}^{(2)} - e_{ij}^{(1)} e_{ij}^{(1)} \right] dV = \int_{\mathcal{D}} \Delta e_{ij} \Delta e_{ij} dV \geqslant 0 \tag{2.49}$$

This quantity is always positive or zero, which means that the Stokes flow (1) dissipates less energy than the general flow (2). Equality occurs only if $\Delta e_{ij} = 0$, or if $\mathbf{u}^{(2)} = \mathbf{u}^{(1)}$.

2.6 Reciprocal Theorem

The reciprocal theorem, although a little abstract and austere, is an essential tool for the study of Stokes flows. It establishes a relationship between two

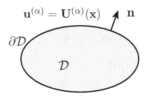

FIGURE 2.2
Definition of the fluid domain. The velocity is prescribed on the boundary.

different Stokes flows in the same domain. Usually, one flow is a known fundamental solution of the Stokes equations and the second one is the complex flow that is sought as part of the solution of a given problem. Its usefulness will be illustrated in different chapters of this book.

2.6.1 Reciprocal Theorem

Let $\mathbf{u}^{(1)}$, $\boldsymbol{\sigma}^{(1)}$ and $\mathbf{u}^{(2)}$, $\boldsymbol{\sigma}^{(2)}$ be the velocity and stress fields in two different fluids with respective viscosity $\mu^{(1)}$ and $\mu^{(2)}$, undergoing Stokes flow in a domain $\mathcal{D}$ with boundary conditions

$$\mathbf{u}^{(1)} = \mathbf{U}^{(1)}(\mathbf{x}) \quad \text{and} \quad \mathbf{u}^{(2)} = \mathbf{U}^{(2)}(\mathbf{x}) \quad \text{for} \quad \mathbf{x} \in \partial \mathcal{D} \tag{2.50}$$

Then

$$\int_{\partial \mathcal{D}} \left[\mu^{(1)} u_i^{(1)} \sigma_{ij}^{(2)} n_j - \mu^{(2)} u_i^{(2)} \sigma_{ij}^{(1)} n_j \right] dS$$

$$= \int_{\mathcal{D}} \left[\mu^{(1)} u_i^{(1)} \frac{\partial \sigma_{ij}^{(2)}}{\partial x_j} - \mu^{(2)} u_i^{(2)} \frac{\partial \sigma_{ij}^{(1)}}{\partial x_j} \right] dV \tag{2.51}$$

where $\mathbf{n}$ is a unit vector normal to ∂D pointing out of $\mathcal{D}$ (Figure 2.2).

2.6.2 Demonstration

We assume that $\partial \sigma_{ij}/\partial x_j$ can take a non-zero value for some singular points in $\mathcal{D}$ (Chapter 8). We derive the product $u_i^{(1)} \sigma_{ij}^{(2)}$ with respect to x_j and use the fact that the fluid is Newtonian

$$\frac{\partial}{\partial x_j} \left[u_i^{(1)} \sigma_{ij}^{(2)} \right] = \frac{\partial u_i^{(1)}}{\partial x_j} \left[-p^{(2)} \delta_{ij} + 2\mu^{(2)} e_{ij}^{(2)} \right] + u_i^{(1)} \frac{\partial \sigma_{ij}^{(2)}}{\partial x_j}$$

$$= 2\mu^{(2)} e_{ij}^{(2)} e_{ij}^{(1)} + u_i^{(1)} \frac{\partial \sigma_{ij}^{(2)}}{\partial x_j} \tag{2.52}$$

The same operation on $u_i^{(2)}\sigma_{ij}^{(1)}$ leads to

$$\frac{\partial}{\partial x_j}\left[u_i^{(2)}\sigma_{ij}^{(1)}\right] = 2\mu^{(1)}e_{ij}^{(1)}e_{ij}^{(2)} + u_i^{(2)}\frac{\partial \sigma_{ij}^{(1)}}{\partial x_j} \qquad (2.53)$$

Combining Equations (2.52) and (2.53), we obtain the local form of the generalised reciprocal theorem

$$\frac{\partial}{\partial x_j}\left[\mu^{(1)}u_i^{(1)}\sigma_{ij}^{(2)}\right] - \frac{\partial}{\partial x_j}\left[\mu^{(2)}u_i^{(2)}\sigma_{ij}^{(1)}\right]$$

$$= \mu^{(1)}u_i^{(1)}\frac{\partial \sigma_{ij}^{(2)}}{\partial x_j} - \mu^{(2)}u_i^{(2)}\frac{\partial \sigma_{ij}^{(1)}}{\partial x_j} \qquad (2.54)$$

Integrating Equation (2.54) over $\mathcal{D}$ and applying Gauss' theorem, we obtain the integral form (2.51) of the reciprocal theorem in a domain $\mathcal{D}$.

2.6.3 Particular Cases of the Reciprocal Theorem

When the two fluids have the same viscosity, Equation (2.54) simplifies to

$$\frac{\partial}{\partial x_j}\left[u_i^{(1)}\sigma_{ij}^{(2)}\right] - \frac{\partial}{\partial x_j}\left[u_i^{(2)}\sigma_{ij}^{(1)}\right] = u_i^{(1)}\frac{\partial \sigma_{ij}^{(2)}}{\partial x_j} - u_i^{(2)}\frac{\partial \sigma_{ij}^{(1)}}{\partial x_j} \qquad (2.55)$$

When there are no singular points inside domain $\mathcal{D}$, the right-hand side of Equation (2.54) is zero:

$$\frac{\partial}{\partial x_j}\left[\mu^{(1)}u_i^{(1)}\sigma_{ij}^{(2)}\right] - \frac{\partial}{\partial x_j}\left[\mu^{(2)}u_i^{(2)}\sigma_{ij}^{(1)}\right] = 0 \qquad (2.56)$$

or in integral form,

$$\int_{\partial \mathcal{D}}\left[\mu^{(1)}u_i^{(1)}\sigma_{ij}^{(2)}n_j - \mu^{(2)}u_i^{(2)}\sigma_{ij}^{(1)}n_j\right] dS = 0 \qquad (2.57)$$

Finally, when the fluids have the same viscosity and when there are no singularities in $\mathcal{D}$, we obtain the simplest form of the reciprocal theorem:

$$\int_{\partial \mathcal{D}}\left[u_i^{(1)}\sigma_{ij}^{(2)}n_j - u_i^{(2)}\sigma_{ij}^{(1)}n_j\right] dS = 0 \qquad (2.58)$$

2.7 Solution in Terms of Harmonic Functions

The pressure Laplace equation (2.6) has a simple fundamental solution

$$p = \frac{1}{r} \quad \text{where} \quad r = (x_k x_k)^{1/2} = \sqrt{x_1^2 + x_2^2 + x_3^2} \qquad (2.59)$$

It is possible to create a new solution by applying the gradient operator to $1/r$

$$\frac{\partial}{\partial x_i}\left(\frac{1}{r}\right) = -\frac{x_i}{r^3} \tag{2.60}$$

Repeating this operation, we obtain a series of fundamental solutions of Equation (2.6)

$$p = \frac{S_0}{r}, \quad -S_i\frac{x_i}{r^3}, \quad S_{ij}\left(\frac{-\delta_{ij}}{r^3} + \frac{3x_ix_j}{r^5}\right), \quad \cdots \quad S_{i_1\ldots i_n}\Phi_{-(n+1)} \tag{2.61}$$

with general term

$$\Phi_{-(n+1)} = \frac{\partial^n\left(\frac{1}{r}\right)}{\partial x_{i_1}\ldots\partial x_{i_n}} \qquad n = 0, 1, 2, \cdots \tag{2.62}$$

where $\partial^0 f = f$ and where S_0, S_i, $\ldots S_{i_1\ldots i_n}$ are constant coefficients. These functions decrease when r increases and are singular for $r = 0$. It is possible to form a series of other solutions that increase with r and are regular for $r = 0$

$$p = A_0, \quad A_i x_i, \quad A_{ij}\left(\frac{-r^2\delta_{ij}}{3} + x_ix_j\right), \quad \cdots \quad A_{i_1\ldots i_n}r^{2n+1}\Phi_{-(n+1)} \tag{2.63}$$

where A_0, A_i, $\ldots A_{i_1\ldots i_n}$ are also constant coefficients. Using the relations obtained in Section 2.1.1, we readily find the velocity field. The advantage of these general solutions is that they are independent of the coordinate system. Using the linearity property, we can seek the unique solution to a Stokes flow problem as the linear combination of Equations (2.62) and (2.63) (for the pressure field) which satisfies the boundary conditions of the specific problem. This procedure will be presented in detail in Chapter 8.

2.8 Problems

2.8.1 Symmetry of Stokes Flow

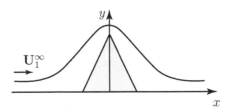

FIGURE 2.3
Stokes flow around a solid particle with a symmetry plane.

A solid body with symmetry plane $x = 0$ is held motionless in a uniform flow with upstream velocity normal to the symmetry plane (Figure 2.3).

1. Show that the streamlines of the flow around the particle are symmetric with respect to plane $x = 0$.
2. Find the direction of the force exerted by the fluid on the solid body

2.8.2 Energy Dissipation Due to the Motion of a Solid Particle in a Quiescent Fluid

1. A rigid particle with surface S moves in a fluid at rest which extends far from the particle. Note that the flow velocity due to the particle motion decreases as $1/r$, where r is the distance from the particle centre (Chapter 6). Show that the energy dissipation Φ is given by

$$\Phi = \int_S \sigma_{ij} u_i n_j \, dS$$

2. Show that the energy dissipation Φ is also given by

$$\Phi = \int_S \sigma_{ij} u_i n_j \, dS$$

when the particle is confined inside a reservoir R with rigid walls.

3. Compute Φ when the particle has a rigid body motion

$$\mathbf{U} = \mathbf{U}_0 + \boldsymbol{\Omega} \times \mathbf{r}$$

where $\mathbf{U}_0$ and $\boldsymbol{\Omega}$ are the translation and rotation velocities of the particle, respectively.

Hint: Consider the energy dissipation in a fluid domain $\mathcal{D}$, bounded by S and by a surface S_∞ which is located far from the particle.

2.8.3   Drag on a Particle in a Reservoir

When a solid particle has a translation motion with velocity $\mathbf{U}_0$ in a fluid with viscosity μ enclosed in a reservoir R_1, it is subjected to a drag force $\mathbf{F}^{(1)}$. The particle is now suspended in the same fluid enclosed in another reservoir R_2 such that $R_2 \subset R_1$. Compare the drag forces $\mathbf{F}^{(1)}$ and $\mathbf{F}^{(2)}$ for the same translation velocity $\mathbf{U}_0$ in the two reservoirs R_1 and R_2, respectively.

3

Two-Dimensional Stokes Flows

CONTENTS

Two-dimensional (2D) flows are flows for which one spatial direction has no influence. The flow occurs in a plane and is identically repeated in parallel planes. This situation is encountered for flows in corners (wedges) or around prismatic obstacles. In such cases, we can frequently neglect end effects and compute the flow near the centre of the obstacle. All the flows studied in this chapter are two-dimensional and correspond to different practical applications, such as knife painting, flow in wedges or around protruding corners, etc.

3.1   Stream Function

3.1.1   Cartesian Coordinates

In order to simplify the presentation, we leave aside the index notation and use a Cartesian reference frame (O, x, y, z). The flow takes place in the xy-

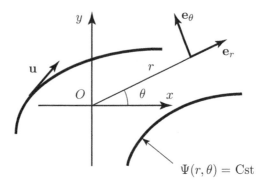

FIGURE 3.1
2D Stokes flow in the xy-plane in terms of Cartesian (x, y) or polar (r, θ) coordinates.

plane and does not depend on the z-direction (Figure 3.1). Consequently, the velocity has components only in the xy-plane which depend only on x and y (and time eventually)

$$\mathbf{u} = u(x, y)\mathbf{e}_x + v(x, y)\mathbf{e}_y \tag{3.1}$$

Similarly, the pressure depends only on x and y (and time eventually)

$$p = p(x, y) \tag{3.2}$$

It is useful to introduce a scalar stream function $\Psi(x, y)$ which allows us to satisfy identically the continuity equation $\nabla.\mathbf{u} = 0$

$$u = \partial\Psi/\partial y \quad \text{and} \quad v = -\partial\Psi/\partial x \tag{3.3}$$

Since a stream line is, by definition, tangent to the velocity at each point, it satisfies

$$\frac{dx}{u} = \frac{dy}{v} \quad \rightarrow \quad -\frac{\partial\Psi}{\partial x}dx = \frac{\partial\Psi}{\partial y}dy \tag{3.4}$$

or

$$d\Psi = 0 \quad \Rightarrow \quad \Psi = \text{Cst} \tag{3.5}$$

We thus find that the stream function is constant along stream lines.

The flow vorticity $\boldsymbol{\omega}$ has only one non-zero component directed along Oz

$$\boldsymbol{\omega} = \nabla \times \mathbf{u} = \left(\frac{\partial v}{\partial x} - \frac{\partial u}{\partial y}\right)\mathbf{e}_z = -\left(\frac{\partial^2\Psi}{\partial x^2} + \frac{\partial^2\Psi}{\partial y^2}\right)\mathbf{e}_z = -\mathfrak{D}^2\Psi\mathbf{e}_z \tag{3.6}$$

where the 2D Stokes operator $\mathfrak{D}^2$ is identical to the 2D Laplacian operator

$$\mathfrak{D}^2\Psi = \left(\frac{\partial^2}{\partial x^2} + \frac{\partial^2}{\partial y^2}\right)\Psi \tag{3.7}$$

3.1.2 Cylindrical Coordinates

It is often desirable to use cylindrical coordinates (r, θ, z) with axis Oz, which correspond to polar coordinates (r, θ) in planes normal to Oz (Figure 3.1). In this system, the 2D velocity and pressure fields become

$$\mathbf{u} = u_r(r, \theta)\, \mathbf{e}_r + u_\theta(r, \theta)\, \mathbf{e}_\theta \quad \text{and} \quad p = p(r, \theta) \tag{3.8}$$

The stream function is then defined by

$$u_r = \frac{1}{r} \frac{\partial \Psi}{\partial \theta} \quad \text{and} \quad u_\theta = -\frac{\partial \Psi}{\partial r} \tag{3.9}$$

This allows us to satisfy identically the continuity equation which reads in polar coordinates (Appendix B.1)

$$\nabla \cdot \mathbf{u} = \frac{1}{r} \frac{\partial(r u_r)}{\partial r} + \frac{1}{r} \frac{\partial u_\theta}{\partial \theta} = 0 \tag{3.10}$$

The only non-zero component of the vorticity is oriented along Oz

$$\boldsymbol{\omega} = \nabla \times \mathbf{u} = \omega(r, \theta)\, \mathbf{e}_z \tag{3.11}$$

with

$$\omega = \frac{1}{r} \frac{\partial(r u_\theta)}{\partial r} - \frac{1}{r} \frac{\partial u_r}{\partial \theta} = -\mathfrak{D}^2 \Psi \tag{3.12}$$

where the 2D Stokes operator $\mathfrak{D}^2$ is now defined by

$$\mathfrak{D}^2 \Psi = \frac{1}{r} \frac{\partial}{\partial r}\left(r \frac{\partial \Psi}{\partial r} \right) + \frac{1}{r^2} \frac{\partial^2 \Psi}{\partial \theta^2} \tag{3.13}$$

which is identical to the 2D cylindrical Laplacian operator.

3.2 Two-Dimensional Stokes Momentum Equation

Using Equation (3.12) and the form (2.11) of the Stokes equations, we find the following momentum equation

$$\mathfrak{D}^2 (\mathfrak{D}^2 \Psi) = 0 \tag{3.14}$$

which is the bi-harmonic equation.

In many Stokes flow problems, we are interested only in the flow in the vicinity of a boundary or of a moving body. For example, we may seek the stress distribution on a wall in order to study the resuspension of a sediment or the possible damage on adhering cells. Whatever the flow is far from the wall, the no-slip condition on a stationary wall slows down the fluid. There

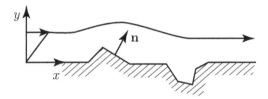

FIGURE 3.2
Stokes flow near a wall with longitudinal obstacles in the z-direction.

is thus a Stokes regime in a small neighbourhood of the wall that must be matched eventually to the outer flow (Figure 3.2). In general, the problem consists of solving Equation (3.14) with a no-slip condition on a wall with velocity $\mathbf{U}$

$$\nabla\Psi \cdot \mathbf{n} = \mathbf{U} \cdot \mathbf{t} \quad \text{on} \quad \partial\mathcal{D} \tag{3.15}$$

$$\nabla\Psi \cdot \mathbf{t} = -\mathbf{U} \cdot \mathbf{n} \quad \text{on} \quad \partial\mathcal{D} \tag{3.16}$$

where $\mathbf{n}$ and $\mathbf{t}$ denote the unit vectors normal and tangent to $\partial\mathcal{D}$, respectively. We must also add a matching condition with the outer flow. In the general case of a boundary with a complex geometry, it is very difficult to obtain an exact analytical solution of Equation (3.14) and we have to resort to numerical solutions (Chapter 8).

In some simple situations, it is possible to find local asymptotic solutions where the variables are separable. For example, flows in corners constitute a class of problems for which self-similar solutions can be found [41]. Using a polar coordinate system centred on the corner apex, we seek a solution of the form

$$\Psi = r^\lambda f(\theta) \tag{3.17}$$

where λ is a complex eigenvalue of the problem which depends on geometry. For the velocity to remain finite, the real part of λ must be larger than unity

$$\Re(\lambda) > 1 \tag{3.18}$$

The solution procedure is illustrated on some examples as follows.

3.3   Wedge with a Moving Boundary: Taylor Paint-Scraper

The first situation we consider is encountered when spreading a Newtonian liquid of viscosity μ on a flat plate. The coating is performed with a plane scraper making an angle α with the plate and moving with velocity $\mathbf{U} = U\,\mathbf{e}_x$

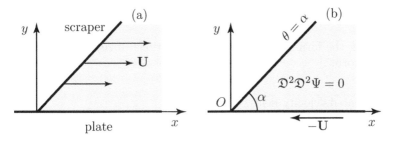

FIGURE 3.3
The scraper makes an angle α with the plate and moves with velocity $\mathbf{U} = U\,\mathbf{e}_x$ with respect to the plate: (a) laboratory reference frame; (b) reference frame moving with the scraper.

(Figure 3.3a). The fluid is contained in the wedge formed by the plate and the scraper. We assume that the edge effects along Oz can be neglected and that the flow is two-dimensional in the xy-plane. The objective is to compute the flow in the corner in order to determine the relation between the force on the scraper and the thickness of the thin liquid film which is deposited on the plate.

This problem was first considered and solved by Taylor [53] and is known as the 'Taylor paint scraper' problem. Of course, apart from the artistic application of knife painting, this flow problem is also encountered in more practical situations: buttering a toast, plastering a wall, etc.

3.3.1 Velocity Field

If we use the laboratory reference frame, the flow domain varies in time because the scraper moves. It is convenient to use a reference frame linked to the scraper and thus to make the flow domain stationary. The scraper is then motionless and the plate moves with velocity $-U\,\mathbf{e}_x$ (Figure 3.3b). We use polar coordinates (r, θ), centred on the intersection O of the plate and the scraper in the plane of flow. The velocity and pressure fields are given by Equation (3.8). The problem thus consists of solving the bi-harmonic equation (3.14) with boundary conditions

$$u_r(r,0) = \frac{1}{r}\frac{\partial\Psi}{\partial\theta} = -U \quad \text{and} \quad u_\theta(r,0) = -\frac{\partial\Psi}{\partial r} = 0 \quad \text{for} \quad \theta = 0$$

$$u_r(r,\alpha) = 0 \quad \text{and} \quad u_\theta(r,\alpha) = 0 \quad \text{for} \quad \theta = \alpha \qquad (3.19)$$

Equations (3.19) can be integrated with respect to r:

$$\frac{\partial\Psi}{\partial\theta} = -Ur, \quad \Psi = 0 \quad \text{for} \quad \theta = 0$$

$$\frac{\partial \Psi}{\partial \theta} = 0, \quad \Psi = 0 \quad \text{for} \quad \theta = \alpha \tag{3.20}$$

Considering the form of the boundary conditions (3.20), we seek a solution under the form

$$\Psi(r, \theta) = U r \, f(\theta) \tag{3.21}$$

with associated velocity field

$$u_r(r, \theta) = U f'(\theta), \qquad u_\theta(r, \theta) = -U f(\theta) \tag{3.22}$$

Conditions (3.20) become

$$f(0) = 0, \quad f'(0) = -1, \quad f(\alpha) = 0, \quad f'(\alpha) = 0 \tag{3.23}$$

where $f' = df/d\theta$.

The solution of the bi-harmonic equation is done in two steps. First, we compute

$$\mathfrak{D}^2 \Psi = \frac{1}{r} \frac{\partial}{\partial r} \left(r \frac{\partial \Psi}{\partial r} \right) + \frac{1}{r^2} \frac{\partial^2 \Psi}{\partial \theta^2} = \frac{U}{r} [f(\theta) + f''(\theta)] \tag{3.24}$$

and thus find

$$\mathfrak{D}^2 \Psi = \frac{U}{r} F(\theta) \tag{3.25}$$

where the auxilliary function $F(\theta)$ is defined by

$$F(\theta) = f(\theta) + f''(\theta) \tag{3.26}$$

We apply again the bi-harmonic operator to Equation (3.25):

$$\mathfrak{D}^2 \mathfrak{D}^2 \Psi = \mathfrak{D}^2 \left[\frac{U}{r} F(\theta) \right] = \frac{U}{r^3} (F + F'') = 0 \tag{3.27}$$

The solution of Equation (3.27) is simple:

$$F(\theta) = A' \cos \theta + B' \sin \theta \tag{3.28}$$

The value of $f(\theta)$ is obtained after the integration of Equation (3.26)

$$f(\theta) = A \cos \theta + B \sin \theta + C\theta \cos \theta + D\theta \sin \theta$$
$$\text{with} \quad C = -B'/2 \quad \text{and} \quad D = A'/2 \tag{3.29}$$

The boundary conditions (3.23) allow us to determine the coefficients A, B, C and D and to find the expression of $f(\theta)$:

$$f(\theta) = \frac{\alpha(\alpha - \theta) \sin \theta - \theta \sin(\alpha - \theta) \sin \alpha}{\sin^2 \alpha - \alpha^2} \tag{3.30}$$

Note that there is no specified length scale in the problem so that we have only computed the flow in the 'vicinity' of point O. A Reynolds number is then defined, based on the distance from O:

$$Re = Ur/\nu \qquad (3.31)$$

This Reynolds number increases with r (we have here a situation which is similar to the one encountered in the study of the boundary layer on a flat plate where the Reynolds number is based on the distance along the plate). The corresponding streamlines are shown for $\alpha = \pi/2$ in Figure 3.4. The approximate solution to the Stokes equations is compared to a numerical solution of the Navier-Stokes equations where inertia terms are taken into account even though they are small [26]. We note that as long as $Re < 2$, the Stokes solution is very near the exact solution. This may seem surprising since the Stokes solution is supposed to be valid only for $Re \ll 1$. In fact, it turns out that $O(1)$ values of the Reynolds numbers are still small enough for inertia effects to be negligible.

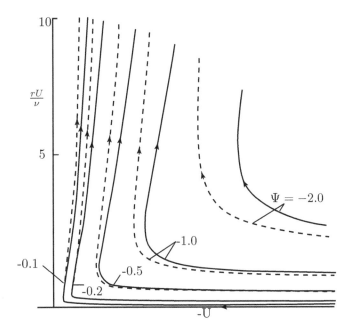

FIGURE 3.4

Sketch of the streamlines for different values of Ψ and comparison with an exact solution for $\alpha = \pi/2$. Full line: approximate Stokes equation; dashed line: solution of the Navier-Stokes equations. (Reproduced from Hancock, Lewis and Moffatt [26] with permission from Cambridge University Press.)

3.3.2   Pressure Field

In order to evaluate the forces exerted by the fluid on the scraper, we must compute the stress field in the fluid. The viscous part of the stress tensor is determined from the velocity field. But the pressure in the fluid must also be computed. We start with the form (2.14) of the Stokes equation that relates the pressure to the vorticity. The radial component of Equation (2.14) is

$$\frac{\partial p}{\partial r} = -\mu \frac{1}{r} \frac{\partial \omega}{\partial \theta} \tag{3.32}$$

Using Equations (3.12) and (3.25), we find

$$\frac{\partial p}{\partial r} = -\frac{\mu}{r} \frac{\partial}{\partial \theta} \left(-\mathfrak{D}^2 \Psi \right) = \frac{\mu}{r} \frac{\partial}{\partial \theta} \left(\frac{U}{r} F(\theta) \right) \tag{3.33}$$

which is easy to integrate:

$$p(r, \theta) - p_0 = -\frac{\mu U}{r} F'(\theta) = \frac{2\mu U}{r} \frac{\alpha \sin\theta + \sin(\alpha - \theta) \sin\alpha}{\alpha^2 - \sin^2\alpha} \tag{3.34}$$

We thus find an infinite pressure for $r = 0$, which is physically impossible! This singularity is due to the velocity boundary condition at point O. Indeed at O, the plate velocity is $-U\,\mathbf{e}_x$, whereas it is zero on the scraper. This means that the plate and the scraper cannot be in contact at O and that a very thin liquid film flows under the scraper. This also means that the solution we have derived is approximate as it breaks down in the corner.

3.3.3   Force on the Scraper

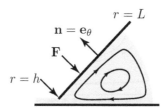

FIGURE 3.5
Forces on the scraper and on the fluid. The streamline pattern is also shown.

The force per unit surface area exerted by the scraper on the fluid is given by $\boldsymbol{\sigma} \cdot \mathbf{n}$, where $\mathbf{n} = \mathbf{e}_\theta$ is the outer unit normal vector to the fluid domain on the scraper (Figure 3.5). The stress tensor is computed from Newton's law,

$$\boldsymbol{\sigma} = -p\mathbf{I} + 2\mu\mathbf{e} = -p\mathbf{I} + \boldsymbol{\tau} \tag{3.35}$$

where τ is the viscous contribution to the stress, which is computed from the velocity field. We thus find

$$\boldsymbol{\sigma} \cdot \mathbf{n} = \tau_{r\theta} \mathbf{e}_r + (-p + \tau_{\theta\theta}) \mathbf{e}_\theta \tag{3.36}$$

It is easy to check that $\tau_{\theta\theta} = 0$ everywhere. For $\theta = \alpha$, a straightforward computation leads to

$$\tau_{r\theta}(r, \alpha) = \frac{2\mu U}{r} \frac{\sin\alpha - \alpha\cos\alpha}{\sin^2\alpha - \alpha^2} \tag{3.37}$$

which shows that the shear stress is also singular near the corner.

Let $\mathbf{F} = F\mathbf{e}_\theta$ be the normal force per unit width of scraper (measured along the z-direction) that we must exert to move the scraper. This force is given by

$$F = \int_h^L -[p(r, \alpha) - p_0]\, dr \tag{3.38}$$

where h is the thickness of the film which flows under the scraper and which is deposited on the plate. The wetted length of scraper is denoted L. Using Equation (3.35), we find

$$F = -2\mu U \ln(L/h) \frac{\alpha \sin\alpha}{\alpha^2 - \sin^2\alpha} \tag{3.39}$$

We can then deduce the film thickness as a function of the exerted force

$$h/L = \exp\left(\frac{-|F|(\alpha^2 - \sin^2\alpha)}{2\mu U \alpha \sin\alpha}\right) \tag{3.40}$$

We find the well-known phenomena occurring when we butter a toast: the thickness of the film decreases when we increase the force on the knife, but increases with the viscosity of the fluid (butter just out of the fridge) or the velocity of the knife. Note that the force also has a non-zero tangential component along the scraper. This solution is, however, approximate because we have not taken into account the flow under the scraper.

3.3.4 Other Wedge Flows with Moving Boundary

Another situation which is very similar to that of the Taylor scraper is when a belt is moving into (or out of) a quiescent liquid as shown in Figure 3.6. The belt makes an angle α with respect to the horizontal direction and plunges into the fluid with velocity U. We seek the flow in the corner formed by the belt at $\theta = -\alpha$ and the free surface of the fluid at $\theta = 0$. The problem boundary conditions are

$$u_r(r, -\alpha) = U, \quad u_\theta(r, -\alpha) = 0, \quad u_\theta(r, 0) = 0, \quad \sigma_{r\theta}(r, 0) = 0 \tag{3.41}$$

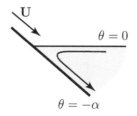

FIGURE 3.6
Belt plunging into a quiescent liquid.

We seek again a solution under the form $\Psi = Urf(\theta)$ which satisfies the bi-harmonic equation (3.14) with conditions

$$f(-\alpha) = 0, \quad f'(-\alpha) = 1, \quad f(0) = f''(0) = 0 \qquad (3.42)$$

We find

$$\Psi = Ur\frac{\theta\cos\theta\sin\theta - \alpha\sin\theta}{\sin\alpha\cos\alpha - \alpha} \qquad (3.43)$$

It can be easily verified that the fluid velocity on the free surface is constant and less than U. The fluid particles must then accelerate when they get near the belt in order to attain the velocity U. Consequently, the acceleration and pressure go to infinity near the corner, which again is not realistic. In fact other phenomena such as capillary and contact forces become important in the corner of the wedge and tend to deform the free surface.

Another situation with moving wall occurs during the opening and closing of a wedge (see Problem 3.5.1).

3.4 Flow in Fixed Wedges

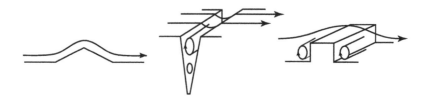

FIGURE 3.7
Examples of flow near 2D grooves or protrusions.

A wide class of 2D Stokes flows deals with the flows near protrusions or grooves in a motionless wall (Figure 3.7). The exact nature of the far-field flow is not important since we are only interested in the phenomena that occur in the vicinity of the wall where the Reynolds number is small. One application of these flows deals with the hydrodynamic cleaning of reaction tanks where polluting substances or bacteria may remain trapped in small cracks or defects of the vessel.

We first seek the flow in the vicinity of the apex of a wedge of angle 2α without concerning ourselves with the actual cause of the flow, as depicted in Figure 3.8. We note that there is no defined length scale for this situation either. This class of problems has been systematically explored by Moffatt [41].

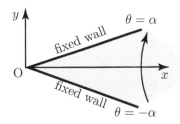

FIGURE 3.8
Flow in a fixed wedge of angle 2α. The far flow is undefined and we only consider the flow near the apex of the wedge.

We use polar coordinates in the xy-plane and the notations of Section 3.3. The boundary conditions are

$$u_r(r, \theta) = 0 \quad \text{and} \quad u_\theta(r, \theta) = 0 \quad \text{for} \quad \theta = \pm\alpha \tag{3.44}$$

As shown by Moffatt [41], we can find a universal solution near $r = 0$ of the the form

$$\Psi(r, \theta) = r^\lambda \, f(\theta), \quad \Re(\lambda) > 1 \tag{3.45}$$

The velocity components are then

$$u_r = \frac{\partial \Psi}{r \partial \theta} = r^{\lambda-1} f'(\theta) \quad \text{and} \quad u_\theta = -\frac{\partial \Psi}{\partial r} = -\lambda r^{\lambda-1} f(\theta) \tag{3.46}$$

The boundary conditions (3.44) become

$$f(\theta) = 0 \quad \text{and} \quad f'(\theta) = 0 \quad \text{for} \quad \theta = \pm\alpha \tag{3.47}$$

In order to solve the bi-harmonic equation, we first apply the Stokes operator to Ψ:

$$\mathfrak{D}^2 \Psi = r^{\lambda-2} [f''(\theta) + \lambda^2 f(\theta)] \tag{3.48}$$

We let $F(\theta) = f'' + \lambda^2 f$, and apply again the operator $\mathfrak{D}^2$:

$$\mathfrak{D}^2 \mathfrak{D}^2 \Psi = \mathfrak{D}^2 \left[r^{\lambda-2} F(\theta) \right] \tag{3.49}$$

which becomes

$$\mathfrak{D}^2 \mathfrak{D}^2 \Psi = r^{\lambda-4} \left[F'' + (\lambda - 2)^2 F \right] = 0 \tag{3.50}$$

with general solution

$$f(\theta) = A \cos(\lambda - 2)\theta + B \cos \lambda\theta + C \sin(\lambda - 2)\theta + D \sin \lambda\theta \tag{3.51}$$

We first seek an antisymmetric solution such that

$$u_r(r, -\theta) = -u_r(r, \theta) \quad \text{and} \quad u_\theta(r, -\theta) = u_\theta(r, \theta) \tag{3.52}$$

In this case, $f(\theta)$ is an even function of θ,

$$f(\theta) = A \cos(\lambda - 2)\theta + B \cos \lambda\theta \tag{3.53}$$

The boundary condition on $\theta = \alpha$ leads to

$$\begin{aligned} f(\alpha) &= A \cos(\lambda - 2)\alpha + B \cos \lambda\alpha = 0, \\ f'(\alpha) &= -A(\lambda - 2) \sin(\lambda - 2)\alpha - B\lambda \sin \lambda\alpha = 0 \end{aligned} \tag{3.54}$$

A non-trivial solution is obtained when λ satisfies the eigen equation of the system (3.54)

$$\sin \left[2\alpha(\lambda - 1) \right] = -(\lambda - 1) \sin 2\alpha \tag{3.55}$$

Of course, if we seek a symmetric solution such that

$$u_r(r, -\theta) = u_r(r, \theta) \quad \text{and} \quad u_\theta(r, -\theta) = -u_\theta(r, \theta) \tag{3.56}$$

$f(\theta)$ is then an odd function of θ but we find the same Eigen Equation (3.55). The general solution is a linear combination of the symmetric and antisymmetric solutions.

3.4.1   Large Wedge Angle ($2\alpha > 146.3°$)

Equation (3.55) has some obvious solutions. The case $\lambda = 1$, which gave the solution to *moving* wedge walls (Section 3.3), corresponds here to a trivial solution that must be discarded. For the case $\alpha = \pi/2$, corresponding to a flat plate, the solution $\lambda = 2$ is a simple shear flow and the solution $\lambda = 3$ is the flow near a stagnation point (see Problem 3.5.3). For the case $\alpha = \pi$, the solution $\lambda = 3/2$ represents the flow around the end of a flat plate.

The roots of the Eigen Equation (3.55) correspond to the intersection between the line $y = (\lambda - 1)\alpha[\sin 2\alpha/\alpha]$ and the curve $y = \sin(\lambda - 1)2\alpha$ as shown in Figure 3.9. It appears that Equation (3.55) has multiple real solutions for $2\alpha > 146.3°$ and no real solution for $2\alpha < 146.3°$. In any case, for a given value of α, Equation (3.55) must be solved numerically.

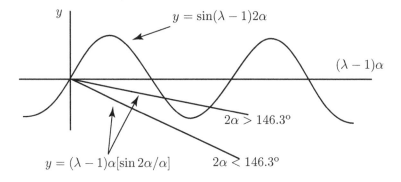

FIGURE 3.9
Graphical representation of Equation (3.55): there is no real solution for $2\alpha < 146.3°$ and multiple real solutions for $2\alpha > 146.3°$.

For $2\alpha > 146.3°$, the dominant term corresponds to the smallest value of λ because the solution has the form r^λ and $r \to 0$. The real roots of Equation (3.55) were tabulated by Dean and Montagnon [19] and are given in graphical form by Pozrikidis [46]. The value of λ decreases from 2.76 for $2\alpha = 146.3°$ to 2 for $2\alpha = 180°$.

The case $\alpha > 73°$ corresponds to a protruding wedge. Typical streamlines are shown for antisymmetric, symmetric and arbitrary flows in Figure 3.10. If we reverse the flow in Figure 3.10(b,c), we depict the case of a jet impinging on a wedge.

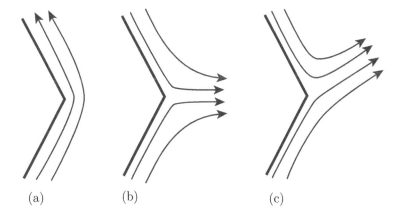

FIGURE 3.10
Streamlines for $\alpha > 73°$: (a) antisymmetric flow; (b) symmetric flow; (c) arbitrary flow.

TABLE 3.1
Complex Solutions of System (3.57) for $\alpha \leq 73.3^\circ$

$\alpha(^\circ)$	0	5	15	25	35	45	55	65	70	73.3
$2\alpha p$	4.21	4.21	4.22	4.24	4.26	4.30	4.35	4.42	4.46	4.48
$2\alpha q$	2.26	2.25	2.20	2.11	1.97	1.77	1.47	1.02	0.64	0

From [41].

3.4.2 Small Wedge Angle $(2\alpha < 146.3^\circ)$

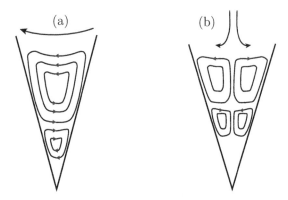

FIGURE 3.11
Streamlines for $\alpha < 73^\circ$: (a) antisymmetric flow; (b) symmetric flow.

When $2\alpha < 146.3^\circ$, the eigen equation (3.55) having no real root, we seek complex solutions under the form $\lambda = (p+1) + iq$. When we separate the real and imaginary parts of Equation (3.55), we obtain two equations for p and q which are also solved numerically

$$\begin{aligned} \sin(2\alpha p)\cosh(2\alpha q) &= -p\sin(2\alpha) \\ \cos(2\alpha p)\sinh(2\alpha q) &= -q\sin(2\alpha) \end{aligned} \tag{3.57}$$

The smallest solutions of the two Equations (3.57) are given in Table 3.1.

In order to understand the flow pattern corresponding to a complex eigenvalue, we compute the tangential velocity u_θ of the fluid:

$$u_\theta(r,\theta) = -\frac{\partial \Psi}{\partial r} = \Re\left\{-\lambda r^{\lambda-1} f(\theta)\right\} \tag{3.58}$$

We thus obtain on the wedge axis $\theta = 0$

$$\begin{aligned} u_\theta(r,0) &= \Re\left\{r^{\lambda-1}K\right\} \quad \text{with} \quad K = -\lambda f(0) = |K|\, e^{i\kappa} \\ u_\theta(r,0) &= r^p\, |K|\, \cos(q\,\ln r + \kappa) \end{aligned} \tag{3.59}$$

When $r \to 0$, $\ln r \to \infty$, and thus $u_\theta(r, 0)$ oscillates about zero. This shows the existence of eddies in the wedge. The centre $r^{(n)}$ of an eddy corresponds to $u_\theta(r^{(n)}, 0) = 0$, and thus to

$$q \ln r^{(n)} + \kappa = -(2n + 1)\,\pi/2 \qquad (3.60)$$

Two successive eddy centres are located at $r^{(n)}$ and $r^{(n+1)}$ such that

$$r^{(n)}/r^{(n+1)} = e^{\pi/q} \qquad (3.61)$$

and the strengths (proportional to the maximum value of u_θ) of two successive eddies are

$$\left| u_\theta^{(n)}/u_\theta^{(n+1)} \right| = e^{\pi\,p/q} \qquad (3.62)$$

For example, for $\alpha = 15°$, the ratio between two successive eddy centres is $r^{(n)}/r^{(n+1)} = 2.1$ with a strength ratio $|u_\theta^{(n)}/u_\theta^{(n+1)}| = 415$, which shows that the eddy intensity dies out quickly as we move near the wedge apex. Antisymmetric eddies are shown in Figure 3.11a. Experiments have been conducted by Taneda [51] in order to visualise these eddies that had been predicted theoretically but never observed. The liquid used is glycerin with small suspended aluminum particles to allow for flow visualisation. Figure 3.12 shows very clearly two successive eddies. Thus, Moffatt's corner eddy theory was confirmed some 15 years after it was first postulated.

These corner eddies are also found in other flows with complex 2D geometry (rectangular grooves, 2D obstacles on a flat wall, etc). For example, depending on the ratio between the width w and the depth h of a rectangular cavity, we observe a single large eddy ($w/h = 2$, Figure 3.13a) or two corner eddies for a wider cavity ($w/h = 3$, Figure 3.13b). Similarly, for protrubing obstacles, eddies also appear in the corners (Figure 3.13c). The numerical model of Higdon [27] also shows the existence of such eddies, in very close accordance with the experimental results (Figure 3.13). Such eddies also occur in the angle between a cylinder and a tangent flat plate or between two tangent cylinders with parallel axes.

These eddies thus appear frequently in 2D flows near walls. Because of the strong dampening of their strength as we get near the apex of the corner, the flow becomes slower and slower, and convective effects decrease accordingly. This phenomenon must be taken into account, in particular for microfluidic systems where this situation arises.

In conclusion, this study of flows in wedges gives a mathematically rigorous analysis of a well-known problem: it is indeed very difficult to clean corners!

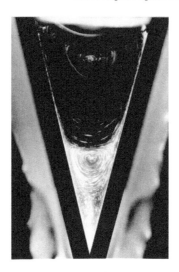

FIGURE 3.12
Experimental observation of corner eddies for $2\alpha = 28.5°$. The fluid is glycerin with suspended aluminium particles. The outer driving flow Reynolds number is $Re = 0.17$, the camera exposition time is 90 min. (Reproduced from Taneda [51] with permission from the Physical Society of Japan.)

(a)   (b)   (c)

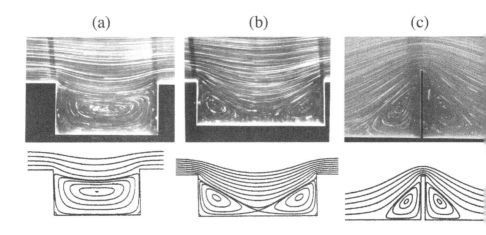

FIGURE 3.13
Corner eddies in cavities with different width w to height h ratios: (a) rectangular cavity $w/h = 2$; (b) rectangular cavity $w/h = 3$; (c) fence. (Top line: experimental observations, reproduced from Taneda [51] with permission from the Physical Society of Japan). Bottom line: numerical simulations, reproduced from Higdon [27] with permission from Cambridge University Press.)

3.5 Problems

3.5.1 Closing a Wedge

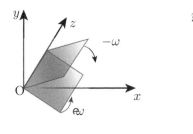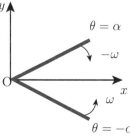

FIGURE 3.14

Closing of a wedge with angle 2α at time t.

Two flat plates rotate about a common axis Oz. The space between the two plates is filled with a Newtonian incompressible fluid (viscosity μ, density ρ). The objective of the problem is to compute the flow in the corner formed by the two plates when the wedge is closed (Figure 3.14). The 'upper' plate (position $\theta = \alpha$ at time t) rotates about the z-axis with angular velocity $-\omega$ ($\omega > 0$). The 'lower' plate (position $\theta = -\alpha$ at time t) rotates about the z-axis with angular velocity $+\omega$. Assume that the flow Reynolds is small enough for the Stokes equations to apply. The plate dimensions along the z-direction are large, so that we can consider that the flow between the two plates is two-dimensional in the xy-plane (Figure 3.14). There are no gravity effects. We use polar coordinates (r, θ) in the xy-plane. The velocity field is thus

$$\mathbf{u} = u_r(r, \theta)\mathbf{e}_r + u_\theta(r, \theta)\mathbf{e}_\theta$$

with associated stream function $\Psi(r, \theta)$.

1. Give the boundary conditions that must be satisfied by the two velocity components $u_r(r, \theta)$ and $u_\theta(r, \theta)$.

2. Seek a solution of the form $\Psi = \omega r^2 g(\theta)/2$. Show that

$$g(\theta) = C \cos 2\theta + D \sin 2\theta + A\theta + B$$

Compute the coefficients A, B, C and D and determine $g(\theta)$.

3. Is the function $g(\theta)$ always determined? Comment.

4. Compute the pressure field $p(r, \theta)$ between the two plates. Comment on the result.

3.5.2   Flow in an Unbounded Cavity

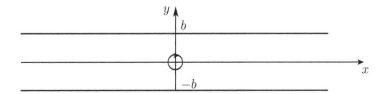

FIGURE 3.15
Flow in an unbounded cavity with parallel walls.

A liquid with viscosity μ is contained inside an infinite cavity made of two parallel plane walls located at $y = \pm b$. A 2D Stokes flow $\mathbf{u} = u(x, y)\mathbf{e}_x + v(x, y)\mathbf{e}_y$ with stream function $\Psi(x, y)$ is created near the origin $x = y = 0$ (for example, by means of a rotating cylinder). We seek the flow far from the origin (x large).

1. Give the boundary conditions for the velocity components.
2. Show that it is possible to find a solution to this problem corresponding to a stream function of the form

$$\Psi = f(y)\exp(-kx), \qquad \Re(k) > 0$$

   where $f(y)$ is an even function of y.
3. Show that k is given by the characteristic equation

$$2kb + \sin 2kb = 0$$

4. Show that this equation has no real roots (except for zero). Determine the flow between the two walls.

3.5.3   Flow near a Stagnation Point on a Plane Wall

From an Ecole Polytechnique problem written with K. Moffatt

It happens that under Stokes flow conditions, a stagnation point may be found in the flow. This occurs, for example, when a slow jet impinges on a solid wall, which is the situation considered in this problem.

We consider the Stokes flow of a viscous incompressible liquid (viscosity μ) in the vicinity of a plane wall located at $y = 0$. We assume that the flow is two-dimensional in the xy-plane and we use polar coordinates (r, θ). We wish to study in detail the flow for which the line $\theta = \alpha$ is a streamline (Figure 3.16). Denoting $\Psi(r, \theta)$ as the stream function of the flow, we seek a solution of the form

$$\Psi = r^3 f(\theta)$$

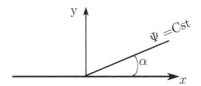

FIGURE 3.16
Flow near a stagnation point on a flat plate. The line $\theta = \alpha$ is a streamline.

1. Find the polar components u_r and u_θ of the velocity. Determine the problem boundary conditions and express them in terms of $f(\theta)$.

2. Find the general expression of $f(\theta)$ which satisfies the Stokes equations.

3. Show that the function $f(\theta)$ which satisfies the problem boundary conditions is given by

$$f(\theta) = K \sin^2 \theta \sin(\theta - \alpha)$$

where K is an arbitrary constant.

Some useful trigonometric formulae

$$\sin 3\theta - 3 \sin \theta = -4 \sin 3\theta \quad \text{and} \quad \cos \theta - \cos 3\theta = 4 \sin 2\theta \cos \theta$$

4. Draw the streamline pattern for $K > 0$ and $\alpha = \pi/4$.

5. Use now Cartesian coordinates (x, y) and show that the pressure gradient ∇p is uniform.

6. Study the limit $\alpha \to 0$. Compute the Cartesian components of the velocity and of the pressure gradient. What is this flow?

4

Lubrification Flows

CONTENTS

This chapter deals with flows confined between two solid walls with a mean separation that is small compared to the distance covered by the fluid. This geometrical constraint leads to a velocity field that is almost parallel to the walls. The viscous effects are then much larger than the inertia effects, and the Stokes equation can be used to describe the flow: this is the so-called *lubrication approximation.*

Lubrication flows are very common. For example, the joint between two bones is lubricated by a thin film of synovial liquid. Similarly, fluid bearings between two mechanical elements in relative motion are lubricated with a thin film of air or oil. Flows in porous media or in microfluidic networks are other examples of confined flows which are dominated by viscous effects and well described by the Stokes equation.

4.1   Two-Dimensional Lubrication Flows

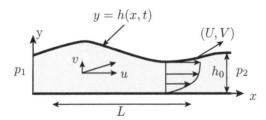

FIGURE 4.1
Two-dimensional lubrication flow between two solid walls.

We first study the case of two-dimensional flows in order to present the different concepts in a simple situation.

We consider the 2D flow of an incompressible liquid (viscosity μ) between two solid walls. The fluid is set into motion by a pressure difference $p_1 - p_2$ across the channel and/or by a relative wall motion. Cartesian coordinates (x, y, z) are used with the flow in the xy-plane as in Chapter 3. The lower wall is flat with equation $y = 0$. The upper wall is irregular with equation $y = h(x, t)$. The mean film thickness between the two walls is denoted h_0. The longitudinal length scale of the walls and of the flow along direction Ox is denoted L (Figure 4.1). The lower wall is fixed and the upper wall has a velocity $\mathbf{U} = U\mathbf{e}_x + V\mathbf{e}_y$. The motion of the upper wall must satisfy the kinematic condition given by Equation (2.17)

$$\frac{D\left[y - h(x, t)\right]}{Dt} = 0 \quad \Rightarrow \quad \frac{\partial h}{\partial t} = V - U\frac{\partial h}{\partial x} \tag{4.1}$$

which links the time variations of h to the wall geometry and velocity.

The parameter $\varepsilon = h_0/L$ measures the aspect ratio of the channel. We assume that two conditions are satisfied:

$$\varepsilon = h_0/L \ll 1 \quad \text{and} \quad |\partial h/\partial x| = O(\varepsilon) \tag{4.2}$$

These two conditions mean that the channel is thin compared to its length and that h varies slowly with x.

4.1.1   Orders of Magnitude and Approximations

The order of magnitude of the flow variables are evaluated using L and h_0 as the length scales along Ox and Oy, respectively. Since the flow is two-dimensional, it is possible to introduce a stream function Ψ (Chapter 3,

Section 3.1. The order of magnitude of the derivatives of Ψ are then

$$|\partial\Psi/\partial y| \sim |\Psi|/h_0 \quad \text{and} \quad |\partial\Psi/\partial x| \sim |\Psi|/L \qquad (4.3)$$

which leads to

$$|\partial\Psi/\partial x| \sim \varepsilon \, |\partial\Psi/\partial y| \qquad (4.4)$$

Since the derivative with respect to x is much smaller than the derivative with respect to y, we find that to first order in ε,

$$\nabla^2\Psi \cong \frac{\partial^2\Psi}{\partial y^2} \qquad (4.5)$$

The relations between Ψ, u and v allow us to evaluate the order of magnitude of the velocities:

$$|v| \sim |\Psi|/L \quad \text{and} \quad |u| \sim |\Psi|/h_0 \qquad (4.6)$$

Thus

$$|v| \sim \varepsilon \, |u| \qquad (4.7)$$

If we scale u and v by $|U|$ and $|V|$, respectively, the relation (4.7) implies that the same relation must exist between the orders of magnitude of the wall velocity components

$$|V| \sim \varepsilon \, |U| \qquad (4.8)$$

The flow field is thus *quasi unidirectional* along direction Ox. It is now possible to evaluate the order of magnitude of the different terms in the Navier–Stokes equations:

$$\begin{aligned} \text{inertia}: \quad & \rho\,|\mathbf{u}\cdot\nabla\mathbf{u}| \sim \rho|U|^2/L \\ \text{viscosity}: \quad & \mu\,|\nabla^2\mathbf{u}| \sim \mu|U|/h_0^2 \end{aligned} \qquad (4.9)$$

The ratio between inertia and viscous terms is thus given by

$$\frac{\rho\,|\mathbf{u}\cdot\nabla\mathbf{u}|}{\mu\,|\nabla^2\mathbf{u}|} \sim \frac{h_0}{L}\frac{\rho|U|h_0}{\mu} = \varepsilon\,Re \qquad (4.10)$$

where the flow Reynolds number is $Re = |U|h_0/\nu$.
The condition

$$\varepsilon Re \ll 1$$

is known as the *lubrication hypothesis*. When it is satisfied, it is possible to use the Stokes equations to describe the flow between the two walls. It is important to note that the lubrication hypothesis does not impose that the Reynolds number be very small. It is essential though that the film aspect ratio be very small, $h_0/L \ll 1$.

4.1.2 Velocity Field

With the above simplifications, the Stokes momentum equation along the x-direction becomes

$$\frac{\partial^2 u}{\partial y^2} = \frac{1}{\mu} \frac{\partial p}{\partial x} \tag{4.11}$$

It follows that the order of magnitude of the pressure is

$$|p| \sim \frac{\mu |U| L}{h_0^2} \tag{4.12}$$

The Stokes momentum equation along the y-direction is

$$\frac{\partial^2 v}{\partial y^2} = \frac{1}{\mu} \frac{\partial p}{\partial y} \tag{4.13}$$

However, the pressure gradient magnitude $|\partial p / \partial y| \sim \mu |U| L / h_0^3$ (obtained from Equation (4.13)) is much larger than the magnitude of the viscous term $|\mu \partial^2 v / \partial y^2| = \mu \varepsilon |U| / h_0^2$. Consequently, the equation of motion along the y-direction is simply

$$\frac{\partial p}{\partial y} = 0 \quad \Rightarrow \quad p = p(x, t) \tag{4.14}$$

This shows that the pressure is constant in a cross section of the film.

The velocity field is computed from Equation (4.11)

$$\frac{\partial^2 u}{\partial y^2} = -\frac{1}{\mu} G(x, t) \tag{4.15}$$

where $G(x, t) = -\partial p / \partial x$ is the pressure gradient in the flow direction. The associated boundary conditions are

$$u = 0 \quad \text{at} \quad y = 0, \quad \text{and} \quad u = U \quad \text{at} \quad y = h(x, t) \tag{4.16}$$

The solution of Equation (4.15) is then

$$u(x, y, t) = -\frac{G(x, t)}{2\mu} y[y - h(x, t)] + U \frac{y}{h(x, t)} \tag{4.17}$$

We thus obtain a Poiseuille–Couette velocity profile between the two walls. The quadratic part (Poiseuille) is due to the pressure gradient, while the linear part (Couette) is due to the wall velocity. There is no contribution from the upper wall velocity V. Indeed, $|V|$ being small with respect to $|U|$ (Equation (4.7)), it is negligible to first order in ε.

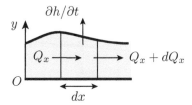

FIGURE 4.2
Mass balance over a small film element.

4.1.3 Mass Balance

The flow rate Q_x per unit width along Oz is defined by

$$Q_x = \int_0^{h(x,t)} u(x,y,t)\, dy = h(x,t)\, \bar{u}_x \qquad (4.18)$$

where $\bar{u}_x$ is the mean velocity in the x-direction. Using Equation (4.17), we find

$$Q_x = \frac{G(x,t)\, h^3}{12\mu} + \frac{U\, h}{2} \qquad (4.19)$$

If we perform a mass balance in a small film element with length dx and unit width along the z-direction (Figure 4.2), we find

$$(Q_x + dQ_x) - Q_x + \frac{\partial}{\partial t}(h\, dx) = 0 \qquad (4.20)$$

When $dx \to 0$, mass conservation requires

$$\frac{\partial h}{\partial t} = -\frac{\partial Q_x}{\partial x} \qquad (4.21)$$

4.1.4 Two-Dimensional Reynolds Equation

Replacing Q_x by Equation (4.19) in Equation (4.21), we obtain the two-dimensional Reynolds equation

$$\frac{\partial}{\partial x}\left[h^3 \frac{\partial p}{\partial x}\right] = 6\mu\left[h\frac{\partial U}{\partial x} + U\frac{\partial h}{\partial x} + 2\frac{\partial h}{\partial t}\right] \qquad (4.22)$$

This is an elliptical differential equation which allows us to compute $p(x,t)$ when $h(x,t)$ and $U(x,t)$ are given. Typical boundary conditions are

$$p = p_1 \quad \text{for} \quad x = x_1 \quad \text{and} \quad p = p_2 \quad \text{for} \quad x = x_2$$

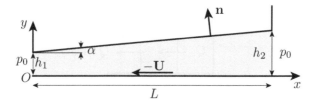

FIGURE 4.3
Two-dimensional slider bearing: the angle α and the gap width between the top and bottom elements are small ($\alpha \ll 1; h_2/L \ll 1$).

4.1.5   Example: Slider Bearing

A slider bearing facilitates the relative motion between two solid mechanical parts by means of a thin liquid film. The simplest bearing has two flat components. The slider is along the xz-plane and the pad makes a small angle $\alpha > 0$ with the xz-plane ($\alpha \ll 1$). Typical values are $\alpha = 10^{-3}$ radians for industrial bearings. The length of the slider along the x-direction is L. The gap width $h(x)$ between the two planes is

$$h(x) = h_1 + \alpha x \qquad \text{with} \qquad \alpha = \frac{h_2 - h_1}{L} \ll 1$$

where h_1 and h_2 are the gap heights at $x = 0$ and $x = L$, respectively (Figure 4.3). We further assume that the film between the pad and slider is thin:

$$\varepsilon = h_2/L \ll 1$$

Finally, we consider the pad and slider as infinite in the z-direction ($L' \gg h_0$, where L' is the width of the slider along the z-direction), so that the flow is two-dimensional in the xy-plane. The gap is filled with a Newtonian incompressible liquid (viscosity μ, density ρ). The outside pressure p_0 is uniform. We use a reference system linked to the pad with respect to which the slider has a constant velocity $-U\mathbf{e}_x$ ($U > 0$). We assume that the lubrication hypothesis is satisfied:

$$\varepsilon \rho U h_2/\mu \ll 1$$

The flow is thus quasi unidirectional in the film:

$$\mathbf{u} = u(x, y)\mathbf{e}_x$$

The solution of Equation (4.15) with the boundary conditions $u(x, 0) = -U$ and $u[x, h(x)] = 0$ yields to first order

$$u(x, y) = -\frac{1}{2\mu}G(x)y[y - h(x)] - U\frac{h(x) - y}{h(x)} \qquad (4.23)$$

which is is similar to Equation (4.17) but is expressed in a different reference system. The flow rate Q_x is constant and given by

$$Q_x = \frac{G(x)h^3(x)}{12\mu} - \frac{Uh(x)}{2}$$

from which we obtain the pressure gradient

$$-G(x) = \frac{dp}{dx} = -\frac{12\mu Q_x}{h^3(x)} - \frac{6\mu U}{h^2(x)} \tag{4.24}$$

For given pressure boundary conditions, the integration of Equation (4.24) leads to the pressure distribution in the gap and to the value of Q_x. For example, for $p(0) = p(L) = p_0$, the pressure in the gap is given by

$$p = p_0 + \frac{6\mu Q_x}{\alpha}\left[\frac{1}{h^2(x)} - \frac{1}{h_1^2}\right] + \frac{6\mu U}{\alpha}\left[\frac{1}{h(x)} - \frac{1}{h_1}\right] \tag{4.25}$$

where we have used the condition $p(0) = p_0$. Note that the flow rate Q_x is still unknown. Using now the condition $p(L) = p_0$ in Equation (4.25), we find the flow rate

$$Q_x = -\frac{h_1 h_2}{h_1 + h_2}U$$

and the velocity profile in the gap

$$\frac{u(x,y)}{U} = y[h(x) - y]\left[\frac{-6h_1 h_2}{h^3(x)(h_1 + h_2)} + \frac{3}{h^2(x)}\right] - \frac{h(x) - y}{h(x)} \tag{4.26}$$

The velocity profile and the pressure distribution in the gap are shown in Figure 4.4. We note that the pressure reaches a maximum value for $h = 2h_1 h_2/(h_1 + h_2)$. The fluid flows with the pressure gradient in the narrowest part of the gap and against it in the wider part. It can easily be checked that these profiles depend on the slider bearing geometry.

The force $\mathbf{F}$ per unit width along Oz exerted by the fluid on the pad is

$$\mathbf{F} = -\int_0^L \boldsymbol{\sigma} \cdot \mathbf{n}\, dS$$

where $\mathbf{n} = -\alpha\mathbf{e}_x + \mathbf{e}_y$ is the unit normal vector pointing out of the fluid (Figure 4.3). The force exerted by the fluid on the pad can be decomposed into a friction component F_x and a lift component F_y. Integrating and neglecting terms $O(\alpha^2)$, we find

$$F_y = \int_0^L [p(x) - p_0]\, dS = -\frac{6\mu U L^2}{h_2^2 - h_1^2}\left[2 - \frac{h_1 + h_2}{h_2 - h_1}\ln\left(\frac{h_2}{h_1}\right)\right] \tag{4.27}$$

$$F_x = -\int_0^L [p(x) - p_0]\alpha\, dS + \int_0^L [\alpha\tau_{xx} - \tau_{xy}]\, dS \tag{4.28}$$

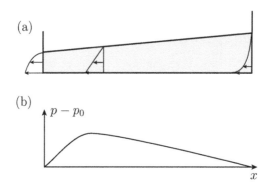

FIGURE 4.4
2D slider bearing: (a) velocity profile in the gap; (b) pressure distribution.
Case $h_2/h_1 = 2$, $h_1/L = 1/10$.

where the viscous contribution τ to the stress tensor is easily computed from
the expression of the velocity field. The friction force F_x becomes

$$F_x = -\frac{18\mu U L}{h_2 + h_1}\left[1 - \frac{4}{9}\frac{h_2 + h_1}{h_2 - h_1}\ln\frac{h_2}{h_1}\right]$$

In the small gap limit $h_2/L \ll 1$, we find

$$|F_y| \cong \frac{12\mu U L^2}{h_2^2 - h_1^2}, \quad |F_x| \cong \frac{18\mu U L}{h_2 + h_1}$$

Note that if $\alpha = 0$, there is no lubrication effect and the pressure p_0 is constant
in the film. Furthermore, there is a lift force on the pad only if the slider is
moved in the proper direction, that is, so that the fluid enters the gap through
the largest section. If we invert the motion of the slider ($U < 0$), the lift force
given by Equation (4.27) becomes negative so that the pad is 'sucked' in by
the fluid. This slider bearing geometry thus works for only one direction of
motion. When we wish to lubricate a two-way periodic motion like the one of
a bone joint (for example, the knee joint), the pad has a cylindrical geometry
with a widening of the gap on each side.

For small angles, the lift force is of order $1/\varepsilon^2$, while the force that must
be exerted to move the bearing is equal and opposite to the friction force
and is thus of order $1/\varepsilon$. This is a very general result that is found for most
lubrication devices: we obtain a large lift force $O(1/\varepsilon^2)$ which is due to pressure
effects in a thin film. The energy necessary to maintain the motion is large
too, $O(1/\varepsilon)$, but is one order of magnitude smaller the the resulting effect.

Slider bearings are used in industry to move heavy loads. For rotational
relative motion, a sector thrust bearing is used where the pad consists of
equally spaced sectors of a circular ring (Figure 4.5).

FIGURE 4.5
Sector thrust bearings used on rotor shafts: example of a Michell thrust bearing with patented pads anchor system. (Courtesy of Meccanica Librandi.)

4.2 Three-Dimensional Lubrication Flows

4.2.1 Three-Dimensional Reynolds Equation

We turn now to 3D flows. The characteristic dimension h_0 measured along the y-direction is assumed to be very small with respect to the smallest dimension L along the x- or z-directions ($\varepsilon = h_0/L \ll 1$). The film thickness $h(x, z, t)$ is now a function of x and z. The velocity of the upper wall is $\mathbf{U} = U\mathbf{e}_x + V\mathbf{e}_y + W\mathbf{e}_z$. The kinematic condition relates the time variations of h to the geometry and to the motion of the wall:

$$\frac{D\left[y - h(x, z, t)\right]}{Dt} = 0 \quad \Rightarrow \quad \frac{\partial h}{\partial t} = V - U\frac{\partial h}{\partial x} - W\frac{\partial h}{\partial z} \qquad (4.29)$$

The length scales are L along the x- and z-directions and h_0 along the y-direction. Similarly, the velocity scales are $|U|, |V|$ and $|W|$ for the velocity components in the x-, y- and z-directions, respectively. We then perform an order of magnitude analysis similar to the one done previously for the 2D case and find that

- The derivatives with respect to x or z are much smaller than the derivative with respect to y

- The flow is quasi two-dimensional and occurs in planes normal to the y-direction

$$\mathbf{u} \cong u(x, y, z, t)\mathbf{e}_x + w(x, y, z, t)\mathbf{e}_z$$

- The pressure is constant in a film cross section

$$\partial p/\partial y = 0, \quad p = p(x, z, t)$$

- The Stokes momentum equations in the x-direction and z-direction become

$$\frac{\partial^2 u}{\partial y^2} = \frac{1}{\mu}\frac{\partial p}{\partial x}, \quad \frac{\partial^2 w}{\partial y^2} = \frac{1}{\mu}\frac{\partial p}{\partial z} \qquad (4.30)$$

- The velocity field in the film is given by

$$
\begin{aligned}
u &= \frac{1}{2\mu}\frac{\partial p}{\partial x}y(y-h) + U\frac{y}{h} \\
w &= \frac{1}{2\mu}\frac{\partial p}{\partial z}y(y-h) + W\frac{y}{h}
\end{aligned}
\qquad (4.31)
$$

Since the Stokes equations are linear, the general solution to the problem is the superposition of two flows in the x- and z-directions. We then define two flow rates Q_x and Q_z per unit width along the z- and x-directions, respectively:

$$Q_x = \int_0^{h(x,z,t)} u(x,y,z,t)\,dy, \qquad Q_z = \int_0^{h(x,z,t)} w(x,y,z,t)\,dy \qquad (4.32)$$

The total flow rate in the film is then $\mathbf{Q} = Q_x\mathbf{e}_x + Q_z\mathbf{e}_z$. Using Equations (4.31) for the velocity field, we find

$$
\begin{aligned}
Q_x &= -\frac{h^3}{12\mu}\frac{\partial p}{\partial x} + U\frac{h}{2} & (4.33) \\
Q_z &= -\frac{h^3}{12\mu}\frac{\partial p}{\partial z} + W\frac{h}{2} & (4.34)
\end{aligned}
$$

or in vector form,

$$\mathbf{Q} = -\frac{h^3}{12\mu}\nabla'p + \mathbf{U}'\frac{h}{2} \qquad (4.35)$$

where $\nabla' = \partial/\partial x\,\mathbf{e}_x + \partial/\partial z\,\mathbf{e}_z$ is the 2D gradient along the x- and z-directions, and where $\mathbf{U}' = U\mathbf{e}_x + W\mathbf{e}_z$. Mass conservation leads to

$$\partial h/\partial t = -\nabla' \cdot \mathbf{Q} \qquad (4.36)$$

and the three-dimensional Reynolds equation writes as

$$\nabla' \cdot [h^3\nabla'p] = 6\mu\left[h\nabla' \cdot \mathbf{U}' + \mathbf{U}' \cdot \nabla'h + 2\frac{\partial h}{\partial t}\right] \qquad (4.37)$$

The 2D or 3D Reynolds equation allows us to compute the pressure distribution between two walls from the knowledge of their geometry and motion, assuming given values of the pressure outside the film. The advantage of the lubrication approximation is that the flows inside and outside the film are decoupled. The pressure inside the film depends only on $h(x,z)$ and on the relative wall motion to first order. The exact value of the external pressure (unless it is $O(1/\varepsilon^2)$) has a marginal influence. For example, in the case of the slider bearing, the pressure in the film can become very large even though there is no pressure gradient between the inlet and outlet of the film. Similarly, the eventual flow outside the film is not affected by the flow inside the film since the flow rate coming in/out the film is very small.

Note that the Reynolds equation is valid only if the lubrication hypothesis is satisfied, that is, if $\varepsilon Re \ll 1$!

4.2.2   Example: Squeezing a Drop

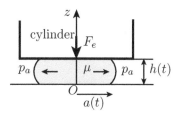

FIGURE 4.6
A volume of liquid squeezed between two parallel plates.

As an example of three-dimensional lubrication flow, we consider a simple axisymmetric situation. A volume V_0 of an incompressible liquid (viscosity μ) is placed between a flat plate and a cylindrical shaft whose end section is parallel to the plate (Figure 4.6). The cylinder has a translation motion with velocity $\mathbf{U} = U\mathbf{e}_z$ parallel to it axis. When the liquid volume is centred, the situation is axisymmetric with revolution axis Oz. We use cylindrical coordinates (r, θ, z) with axis Oz. At time t, the unknown liquid radius is $a(t)$. The pressure outside the liquid is p_a. We neglect surface tension effects on the liquid interface with air. The distance $h(r, t)$ between the plane and the cylinder and the boundary velocity $\mathbf{U}$ are

$$h(r,t) = h(t), \quad \mathbf{U} = \frac{dh}{dt}\mathbf{e}_z \tag{4.38}$$

Since the pressure in the film is constant in a cross section, it depends only on time and radial distance r:

$$p = p(r, t) \tag{4.39}$$

The 3D Reynolds equation (4.37) becomes simply

$$h^3\left[\frac{1}{r}\frac{\partial}{\partial r}\left(r\frac{\partial p}{\partial r}\right)\right] = 12\mu\frac{dh}{dt} \tag{4.40}$$

where the expression of the gradient operator ∇' in cylindrical coordinates is found in Appendix B.1. The associated boundary conditions are

$$p = p_a \quad \text{for} \quad r = a(t)$$

Integrating Equation (4.40) and noting that $p(0,t)$ must be finite, we find the pressure in the film

$$p(r,t) - p_a = \frac{3\mu}{h^3}\frac{dh}{dt}(r^2 - a^2) \tag{4.41}$$

However, we still do not know how fast the liquid spreads. In order to get the time evolution of the radius $a(t)$, we have to perform a force balance on the cylinder. We exert a force $-F_e\mathbf{e}_z$ ($F_e > 0$) to move the cylinder. The axial component of the force exerted by the fluid on the cylinder is

$$F = (-\boldsymbol{\sigma}\cdot\mathbf{e}_z)\cdot\mathbf{e}_z = p - \tau_{zz}$$

It is easy to verify that τ_{zz} is $O(\varepsilon)$ and thus negligible to first order. The force exerted by the fluid on the cylinder is then

$$F = \int_0^a (p - p_a)2\pi r\, dr = -\frac{3\pi\mu}{2h^3}\frac{dh}{dt}a^4 \tag{4.42}$$

Since the fluid is incompressible, the liquid volume V_0 is constant and given by

$$V_0 = \pi a^2 h$$

It is thus possible to write F in terms of V_0:

$$F = -\frac{3\pi\mu}{2h^3}\frac{V_0^2}{\pi^2 h^2}\frac{dh}{dt} = -\frac{3\mu V_0^2}{8\pi}\frac{d}{dt}\left(\frac{1}{h^4}\right) \tag{4.43}$$

The force F is positive and directed upwards only if $dh/dt < 0$, that is, if the film is squeezed. Consequently, the lifting effect occurs only for a given direction of motion (just like for the slider bearing). For a given cylinder velocity, thus for a given value of dh/dt, the pressure in the film and the resultant force F vary like $1/h^4$ and $1/h^5$, respectively. It follows that they increase tremendously as the cylinder gets near the plane. When we neglect inertia forces for the cylinder motion, we find the equilibrium equation:

$$F = F_e \quad \Rightarrow \quad \frac{d}{dt}\left(\frac{1}{h^4}\right) = \frac{8\pi F_e}{3\mu V_0^2} \tag{4.44}$$

which can be integrated

$$h(t) = \left(\frac{3\mu V_0^2}{8\pi F_e}\right)^{1/4}\frac{1}{(t - t_0)^{1/4}} \tag{4.45}$$

where t_0 is a constant. We find a surprising result: the cylinder and the plane will never be in contact! Of course, this cannot be true. In fact, as the film becomes thinner, the roughness and micro defects on the surfaces become important. It follows that the surfaces of the cylinder or of the plane can no longer be treated as 'plane' and the solution we obtained is no longer valid.

Thrust bearings and journal bearings are commonly used for force transmission and for support of rotating shafts. When dealing with such complex devices, the general approach developed above remains valid, even when the cylinder is a rotating shaft. The lubricating liquid is an oil for large mechanical devices such as turbine generator rotor lines. It is air for lighter mechanisms such as a dentist drill or a computer disk drive. In many cases, the large pressure generated in the film requires that the deformation of the wall and the lubricant compressibility be taken into account. Another problem is linked to the mechanical degradation of the lubricating liquid (for large devices). Furthermore, the surface roughness of the different parts becomes an important parameter. The solutions obtained here are thus only valid for films with a thickness much larger than the wall roughness. Finally, in most industrial applications, the lubrication liquid is injected into the film in order to maintain a high enough pressure that will avoid contact between the bearing elements. For air-lubricated bearings, the injection is made through small orifices or though a porous wall.

4.3 Flow between Fixed Solid Boundaries

Lubrication flows also occur frequently between two fixed boundaries, provided that the dimension L along the flow direction is much larger than the cross-section dimension h. The flow velocity is again quasi-2D. This situation is encountered for flows between two parallel plates, in microfluidic channels, in extrusion dies, in porous media, etc.

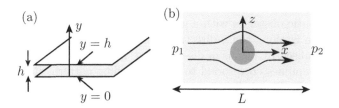

FIGURE 4.7
(a) Hele–Shaw cell; (b) flow around a cylinder, seen from above.

4.3.1 Hele–Shaw Flow

A Hele–Shaw flow cell consists of two fixed parallel plates separated by a gap h, which is much smaller than the plate dimensions (Figure 4.7). The fluid is set in motion by means of a pressure gradient between the entrance and exit sections of the cell. The plates are perpendicular to the y-direction and correspond to planes $y = 0$ and $y = h$, respectively. In the middle of the cell (i.e. far from the lateral sides), the Reynolds equation (4.37) writes as

$$\nabla' \cdot \left[h^3 \nabla' p \right] = 0 \quad \Rightarrow \quad \nabla'^2 p = 0 \tag{4.46}$$

where ∇' is the gradient operator in the xz-plane

$$\nabla' = \frac{\partial}{\partial x} \mathbf{e}_x + \frac{\partial}{\partial y} \mathbf{e}_y \tag{4.47}$$

The flow rate per unit length is then

$$\mathbf{Q} = -\frac{h^3}{12\mu} \nabla' p \tag{4.48}$$

The mean velocity $\bar{\mathbf{u}}$ in the film is defined by

$$\bar{\mathbf{u}} = \bar{u}_x \, \mathbf{e}_x + \bar{u}_z \, \mathbf{e}_z = \mathbf{Q}/h = -\frac{h^2}{12\mu} \nabla' p \tag{4.49}$$

We note that the mean velocity can be interpreted as the velocity of a potential flow with potential function Φ_L

$$\bar{\mathbf{u}} = \nabla' \Phi_L \quad \text{with} \quad \Phi_L = -p \frac{h^2}{12\mu} \tag{4.50}$$

It follows from Equation (4.46) that Φ_L satisfies the Laplace equation

$$\nabla'^2 \Phi_L = 0 \tag{4.51}$$

Hele–Shaw cells can be used to study two-dimensional potential flows around prismatic obstacles with a given cross section and height h. The line of observation is along the y-direction. The streamlines around the obstacle are then visualised by means of die injection. In view of the cell thinness, the observed streamlines correspond to the mean flow pattern. It is amusing to note that a Hele–Shaw cell (where inertia forces are neglected and thus where the Reynolds number is assumed to be very small) can be used to visualise 2D potential flows that correspond to ideal fluid flows where $Re \gg 1$. This is due to the fact that the average velocity (in the Hele–Shaw cell) and the flow velocity (in the 2D potential flow) both derive from a potential function which satisfies the Laplace equation.

For example, we can visualise the streamlines around a cylinder (radius a and height h in the y-direction) subjected to a far-field velocity $\mathbf{U} = U\mathbf{e}_x$ (Figure 4.8). We use polar coordinates (r, θ) centred on the cylinder and we then seek a solution to Equation (4.51) with boundary conditions

$$\begin{aligned} \Phi_L &\to U r \cos \theta \quad \text{for} \quad r \to \infty \\ \partial \Phi_L / \partial r &= 0 \qquad \text{for} \quad r = a \end{aligned} \tag{4.52}$$

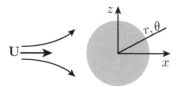
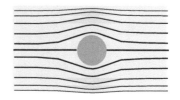

FIGURE 4.8
Flow around a cylinder with axis Oy in a Hele–Shaw flow cell. The flow equations are identical to those those for a 2D inviscid potential flow around a cylinder. Right: streamlines of the flow around a cylinder.

The well-known solution corresponds to the inviscid flow around a cylinder with no circulation:

$$\Phi_L(r, \theta) = U(r + \frac{a^2}{r}) \cos \theta \qquad (4.53)$$

This result is surprising because the inviscid flow has a non-zero slip velocity on the surface of the cylinder. Conversely, the Hele–Shaw flow corresponds to the case where viscous forces are dominant, which implies a no-slip condition on the cylinder surface. So we may ask: where is the catch? In fact, there is no contradiction between the two results, but the reason is rather subtle. The lubrication solution is valid inasmuch the condition $L \gg h$ is satisfied. Near a solid wall (like the cylinder) there is a small region where the distance from the wall is $O(h)$ and where this condition is not satisfied and the lubrication solution not valid. This is why it is not possible to visualise flows with non-zero circulation around a cylinder in a Hele–Shaw cell even if the cylinder is rotating around its axis. Indeed, the fluid vorticity induced by the rotation is killed in a small layer with $O(h)$ thickness around the cylinder.

A Hele–Shaw cell may be considered a very simple model of a confined flow between walls. We can thus use it to study different flows like multiphase flows, relative motion of two immiscible liquids, and such, and treat the results as an ideal approximation of the phenomena that occur in more complex geometries like porous media.

4.3.2 Flow in Microfluidic Channels

Microfluidic systems consist of a network of very small channels (with cross dimension of order or less than 1 mm). The study of flows inside such systems has undergone a booming development in the past decade. The great advantage of these systems is that they use very small fluid volumes with small convection and reaction times. Recent applications of microfluidics are widespread

FIGURE 4.9
Microfluidic system used to cultivate liver cells (hepatocytes) in dynamic flow conditions. The channel width is of order 200 μm. (Courtesy of E. Leclerc, UMR CNRS 7338, Université de Technologie de Compiègne.)

and include parallel screening of molecules with potential therapeutical interest, genomics, chemical or biochemical analysis of liquids for fast detection of toxic components, design of artificial micro-organs based on cell culture in flow (Figure 4.9). Microfluidic channels are usually obtained through soft lithography techniques. A template is first made by etching a wafer of resin of uniform thickness h with the micro-channel network. The template is then moulded in silicone, also called PDMS (polydimethylsiloxane) and glued to a glass plate (for fabrication details, see the review of Xia and Whiteside [55]). As a result, we obtain channels with a rectangular cross section with width w and depth h. It is very difficult to obtain circular channels [50].

The flow in such channels is confined and has a very small Reynolds number as the transversal dimensions of the channel are small. When the channel is straight and has a constant cross section, it is possible to determine the flow field analytically.

As an example, we consider the flow in a microfluidic channel with length L, width w and depth h, such that $h \leq w$ and $(h, w) \ll L$ (Figure 4.10). The flow driving force is a pressure gradient $G = (p_1 - p_2)/L$ applied along the channel axis. Neglecting end effects, the velocity field $\mathbf{u} = u(y, z)\mathbf{e}_x$ is unidirectional along Ox and the pressure $p = p(x)$ is constant in any cross section perpendicular to Ox. We must then solve

$$\frac{\partial^2 u}{\partial y^2} + \frac{\partial^2 u}{\partial z^2} = -\frac{G}{\mu} \qquad (4.54)$$

with boundary conditions

$$u(y, \pm w/2) = 0 \quad \text{and} \quad u(\pm h/2, z) = 0 \qquad (4.55)$$

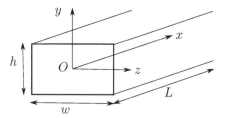

FIGURE 4.10
Microfluidic channel with a rectangular cross section. The flow is along the
x-direction.

We seek a solution in terms of a Fourier series. The constant pressure gradient
can be written as

$$-\frac{G}{\mu} = -\frac{G}{\mu} \frac{4}{\pi} \sum_{n \text{ odd}} \frac{1}{n} \sin n\pi(y/h + 1/2) \tag{4.56}$$

since $\sum_{n \text{ odd}} \frac{1}{n} \sin n\pi(y/h + 1/2) = \pi/4$. Similarly, the velocity field can be expressed as

$$u = \sum_{n \text{ odd}} f_n(z) \sin n\pi(y/h + 1/2) \tag{4.57}$$

which satisfies automatically the no-slip condition on the lateral walls
$y = \pm h/2$. We thus find

$$\frac{\partial^2 u}{\partial y^2} + \frac{\partial^2 u}{\partial z^2} = \sum_{n \text{ odd}} \left(f_n'' - \frac{n^2\pi^2}{h^2} f_n \right) \sin n\pi(y/h + 1/2) \tag{4.58}$$

Equating the right-hand sides of Equations (4.56) and (4.58), we find the
expression for f_n:

$$f_n(z) = \frac{4h^2 G}{\mu\pi^3} \left(\frac{1}{n^3} + A_n \cosh \frac{n\pi z}{h} + B_n \sinh \frac{n\pi z}{h} \right) \tag{4.59}$$

Using the no-slip condition on $z = \pm w/2$, we determine A_n and B_n to finally
obtain the flow field:

$$u(y, z) = \frac{4h^2 G}{\pi^3 \mu} \sum_n \left[\frac{1}{n^3} - \frac{\cosh n\pi z/h}{n^3 \cosh n\pi w/2h} \right] \sin n\pi(y/h + 1/2), \quad n = 1, 3, \ldots \tag{4.60}$$

The flow rate in the channel follows readily:

$$Q = 4 \int_0^{w/2} \int_0^{h/2} u(y, z) \, dy \, dz \tag{4.61}$$

$$Q = \frac{8h^3 wG}{\mu} \left[\frac{1}{96} - \sum_n \frac{\tanh n\pi w/2h}{n^5 \pi^5 w/2h} \right] \qquad n = 1, 3, \ldots \qquad (4.62)$$

The series convergence is fast. Furthermore in Equation (4.62), the hyperbolic tangent varies at most between 0.92 (for a square section) and 1. Consequently, the flow rate is given approximately by

$$Q = \frac{h^3 wG}{12\mu} \left[1 - 0.627 \frac{h}{w} \right] \qquad (4.63)$$

This expression is valid within 1% for $w/h > 1.5$. When $w/h \gg 1$, we retrieve the flow rate in a Hele–Shaw cell (4.48):

$$\mathbf{Q} = -\frac{h^3 w}{12\mu} \nabla' p = \frac{h^3 wG}{12\mu} \qquad (4.64)$$

It should be noted that microfluidic flows can generate high pressures inside the channel. For example, consider a system used for cell culture with $L = 3$ cm, $h = 100\,\mu$m, $w = 200\,\mu$m. Physiological serum ($\mu = 1$ mPa s) which brings nutrients to the cells, flows with a flow rate $Q = 200\,\mu$L/min. This leads to a pressure drop in the microsystem $p_1 - p_2 = 8.7 \times 10^3$ Pa, which is not negligible. If we increase the flow rate or the nutrient fluid viscosity (e.g. by adding sugar), we increase proportionally the pressure inside the microsystem. Independently of its effect on cells, a large internal pressure can deform or damage the system (the connections to the feeding device are particularly fragile).

If the microchannel geometry is more complicated than the one shown in Figure 4.9, we must usually resort to a numerical solution for the internal flow.

4.4   Flow in Porous Media

A porous medium consists of a solid matrix and of pores that are randomly interconnected by microchannels in the general direction of the flow (Figure 4.11). Such media are frequently encountered in nature (porous rocks, granular media, sand, spleen, etc) or in industry (fixed bed, packed catalyst columns, filters, bulk solids, etc). The flow is confined to small narrow tortuous pores with geometry that depends on the general topology of the medium and on the direction of the pressure gradient. In order to control the flow of a fluid through a porous medium, we need to know the relation between pressure drop and flow rate.

FIGURE 4.11
Porous rock.

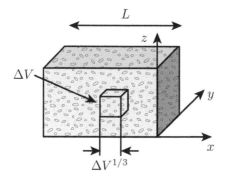

FIGURE 4.12
Characteristic volume of a porous medium.

4.4.1 Definition of a Porous Medium

There are different ways of measuring the geometry of a porous medium. In all cases, the parameters are averaged over a characteristic volume ΔV. The dimension of this volume must be small compared to the macroscopic length L of the medium, but large compared to the dimension d of the pores. This ensures that ΔV contains enough pores for the averaging operation to be meaningful (Figure 4.12):

$$d \ll (\Delta V)^{1/3} \ll L \qquad (4.65)$$

This condition is known as a homogenisation condition and allows us to study a heterogenous medium as a continuum. It will be used again in Chapter 9 for the study of suspensions.

A parameter which is often used is the porosity ϕ_p, defined as the relative interstitial volume occupied by the pores in ΔV,

$$\phi_p = \frac{\text{void volume}}{\Delta V} \tag{4.66}$$

The compactness c_p is the complement of the porosity,

$$c_p = 1 - \phi_p = \frac{\text{solid volume}}{\Delta V} \tag{4.67}$$

The porosity can also be viewed as the relative surface occupied by the void in a cross section of the porous medium. This cross section is arbitrary for an isotropic medium or is normal to the flow direction for an anisotropic medium (e.g. a bundle of parallel capillary tubes - see Problem 4.5.2). However, two media with the same porosity can have pores with different mean size, which leads to different flows. It is thus useful to introduce other parameters, such as the specific internal surface S_p, for example,

$$S_p = \frac{\text{interstitial surface}}{c_p \Delta V} \tag{4.68}$$

We can compute S_p directly if we know the medium geometry. For example, consider a catalyst column consisting of identical spheres (radius a) with a concentration of n_p spheres per unit volume. Then S_p is given by

$$S_p = \frac{n_p 4\pi a^2}{n_p \frac{4}{3}\pi a^3} = \frac{3}{a} \tag{4.69}$$

Note that S_p has the dimension of an inverse length. When the geometry of the medium is random, it is possible to measure S_p by measuring the adsorption of a chemical substance on the internal walls. Another parameter is the tortuosity $T \geq 1$, defined as the length of a typical fluid particle path AB compared to the medium thickness E between the two ends of the path (Figure 4.13)

$$T = AB/E \geqslant 1 \tag{4.70}$$

4.4.2   Flow in a Porous Medium

The dimension d of the pore is usually small enough for the Reynolds number of the flow inside the medium to be small,

$$\frac{Ud}{\nu} \ll 1 \tag{4.71}$$

The local flow thus satisfies the Stokes equations. This means that the pressure drop in a micro channel is proportional to the flow rate, as if the medium consisted of small channels for which Poiseuille's law applied locally. Of course

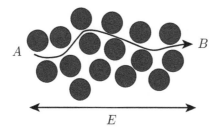

FIGURE 4.13
Definition of the tortuosity.

the proportionality coefficient depends on the channel geometry. Since this geometry is random and impossible to determine with precision, we use mean values of velocity and pressure that are averaged over ΔV:

$$\bar{p} = \frac{1}{\Delta V} \int_{\Delta V} p(x, y, z) dV \tag{4.72}$$

and

$$\bar{u}_x = \frac{1}{\Delta S_x} \int_{\Delta S_x} \mathbf{u}(x, y, z) \cdot \mathbf{e}_x \, dS = \frac{\Delta Q_x}{\Delta S_x} \tag{4.73}$$

where $\bar{u}_x$ is the mean velocity through a section ΔS_x normal to $\mathbf{e}_x$ and where ΔQ_x is the flow rate through this section. The local linear relation between the flow rate and the pressure gradient leads to another linear relation between the average equivalent quantities, which is known as Darcy's law

$$\nabla \bar{p} = -\mu \, \bar{\mathbf{u}} \, / k_p \tag{4.74}$$

where the permeability k_p has the dimension of a surface. The permeability is uniform in a medium which is isotropic and homogeneous. The larger the permeability, the easier it is for a fluid to flow through the porous medium (that is, the larger the flow rate under a given pressure gradient).

The mean velocity $\bar{\mathbf{u}}$ can be related to the velocity potential Φ_L as was done for the analysis of the flow in a Hele–Shaw cell (Equations (4.46) to (4.51)). A typical flow problem in a porous medium then consists of solving the Laplace equation

$$\nabla^2 \Phi_L = 0 \tag{4.75}$$

in a domain $\mathcal{D}$, with boundary condition on $\partial \mathcal{D}$

$$\Phi_L = 0 \quad \text{(free surface)}$$

and /or

$$\partial \Phi_L / \partial n = 0 \quad \text{(impermeable surface)}$$

Darcy's law applies to saturated media that are filled with the same liquid phase. For civil engineering applications, it can be used to compute the flow through the foundations of a dam, the evolution of the water table level, etc. However, there are many situations where two (or more) fluid phases coexist in the porous medium. This is the case of the imbibition of a dry medium by a liquid (percolation of water through ground coffee) or the reverse operation (drying). In another situation, one liquid phase is used to displace another immiscible one (secondary oil extraction, where the oil phase in a porous rock is displaced by an aqueous phase). In such cases, capillarity effects become important at the interfaces between the two phases. They lead to non-linear effects and Darcy's law no longer applies.

4.5 Problems

4.5.1 Lubrication Flow in a Journal Bearing

From an Ecole Polytechnique problem written by F. Dias and K. Moffatt

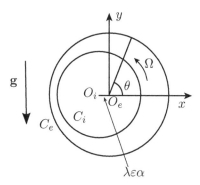

FIGURE 4.14
Journal bearing consisting of two eccentric cylinders, with parallel axes centred on O_i and on O_e for the internal and external cylinder, respectively.

A journal bearing is used to support and guide a rotating shaft. It thus plays an essential role in the set-up of large rotor lines. The objective of the problem is to analyse the fundamental mechanical concepts involved in the design and lubrication of a journal bearing.

A journal bearing consists of a cylindrical ring with radius $a(1+\varepsilon)$, $\varepsilon > 0$, inside which is located a cylindrical shaft with radius a. The gap between the bearing and the shaft is small with respect to the radius a ($\varepsilon \ll 1$). The cylinders' axes $O_i z$ and $O_e z$ are horizontal and separated by $O_i O_e = \lambda \varepsilon a$, $(0 < \lambda < 1)$. The internal cylinder C_i is rotating with angular velocity Ω, while the external cylindrical ring C_e is motionless. The gap between the two cylinders is filled with a Newtonian liquid (viscosity μ). The aim of the problem is to calculate the force exerted on the shaft.

Assume that the bearing longitudinal length is much larger than the typical gap width $a\varepsilon$, so that in the bearing centre, the flow is approximately two-dimensional in a plane normal to the two axes. Furthermore, the film thickness $a\varepsilon$ is much smaller than the film length $2\pi a$, so that the lubrication hypothesis applies to the flow inside the gap. We introduce polar coordinates (r, θ) centred on O_e with corresponding Cartesian coordinates (x, y) where the y-axis is along the vertical upward direction (Figure 4.14).

1. Show that the thickness of the liquid film between the two cylinders

is given by

$$h(\theta) = \varepsilon a(1 - \lambda \cos \theta) + O(\varepsilon^2)$$

2. Show that to leading order in ε

$$\frac{\partial p}{\partial r} = 0, \qquad \frac{1}{a}\frac{\partial p}{\partial \theta} = \mu \frac{\partial^2 u_\theta}{\partial r^2}$$

3. Find the velocity field between the two cylinders in terms of $dp/d\theta$.

4. Compute the flow rate $Q(\theta)$ per unit axial length.

5. Deduce the pressure gradient $dp/d\theta$ as a function of $Q(\theta)$ and integrate it to find finally that the flow rate per unit length Q is given by

$$Q = \Omega a^2 \varepsilon \frac{1 - \lambda^2}{2 + \lambda^2}$$

6. Show that to leading order in ε, the total force $F_y \mathbf{e_y}$ per unit axial length on C_i is given by

$$F_y = -\int_0^{2\pi} p \sin \theta a d\theta = \frac{12\pi\mu\Omega a\lambda}{\varepsilon^2(1 - \lambda^2)^{1/2}(2 + \lambda^2)}$$

7. Find the unique value of λ for which the force F_y balances the weight $-W\mathbf{e}_y$ per unit length of the shaft C_i.

Some useful math...

$$I_n(\lambda) = \int_0^{2\pi} \frac{d\theta}{(1 - \lambda \cos \theta)^n}$$

$$I_1 = \frac{2\pi}{(1 - \lambda^2)^{1/2}}, \qquad I_2 = \frac{2\pi}{(1 - \lambda^2)^{3/2}} \qquad I_3 = \frac{(2 + \lambda^2)\pi}{(1 - \lambda^2)^{5/2}}$$

4.5.2   Flow in an Ideal Porous Membrane

In general, a porous medium consists of random pores with a complicated geometry. Here we study an ideal porous membrane consisting of identical cylindrical parallel pores. We then generalise the concept.

1. A membrane with thickness e is pierced by parallel cylindrical capillary pores with diameter d and length e such that $d \ll e$ (Figure 4.15). There are n pores per unit area of membrane cross section. Compute the permeability k_p in terms of d and of the porosity ϕ_p.

FIGURE 4.15
Ideal membrane made of identical parallel pores with diameter d and length e.

2. For pores with a complex geometry, we introduce V_p and A_p, the volume and the internal surface of each pore, respectively. It is customary to define the pore hydraulic diameter d_h as

$$d_h = \frac{4V_p}{A_p}$$

   Replacing V_p and A_p by their average values, show that for a porous medium with complex pore geometry, the mean hydraulic diameter can be defined by

$$\bar{d}_h = \frac{4\phi_p}{(1-\phi_p)} S_p^{-1}$$

   where S_p is the specific surface of the porous medium.

3. Replace d by $\bar{d}_h$ and write Darcy's law in terms of the fluid viscosity μ, the average flow velocity $\bar{u}_x$, ϕ_p and S_p.

4. In fact, for many porous media it is possible to write Darcy's law as

$$\frac{dp}{dx} = -Ko\frac{\mu\bar{u}_x S_p^2 (1-\phi_p)^2}{\phi_p^3},$$

   where Ko is the so-called Kozeny's constant which is found experimentally to be roughly equal to 5. Compute the permeability k_p of a medium with porosity ϕ_p consisting of spherical particles of diameter d when $Ko = 5$.

4.5.3   Flow along a Porous Wall

From an Ecole Polytechnique problem written with A. Sellier

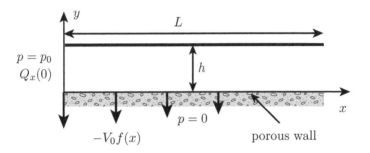

FIGURE 4.16
Lubrication flow along a porous wall with imposed cross flow $v(x,0) = -V_0 f(x)$.

Flowing a suspension of particles or of macromolecules along a porous surface is a separation procedure which is widely used in industry (tangential filtration), in nature (pancreas, kidney) or in medicine (dialysis for kidney disorder). In practice, the pressure difference across the porous membrane leads to the transport of fluid and/or of particles small enough to pass through the pores of the membrane, the larger particles being left in the main flow. The objective of this problem is to study the flow of a pure fluid along a porous membrane in order to analyse the fluid mechanics of the process. The filtration *per se* is not considered in this problem.

A viscous Newtonian liquid (viscosity μ, density ρ) flows between two parallel plates. The flow is two-dimensional in the xy-plane. The plates have length L along Ox and are located at $y = 0$ and $y = h > 0$ with $\varepsilon = h/L \ll 1$ (Figure 4.16). The aim of the problem is to compute the velocity $\mathbf{u} = u(x,y)\mathbf{e_x} + v(x,y)\mathbf{e_y}$ and pressure $p(x,y)$ in the fluid when gravity forces are negligible. We assume that the lubrication hypothesis is satisfied,

$$\varepsilon Re \ll 1 \quad \text{with} \quad Re = \rho U h/\mu$$

where U is the characteristic dimension of u. The upper plate (at $y = h$) is impermeable. However, it is possible to inject or suck fluid across the lower plate (at $y = 0$) according to the law

$$v(x,0) = -V_0 f(x) \qquad (4.76)$$

where V_0 ($V_0 > 0$) and the function $f(x)$ are given. The pressure at the entrance of the channel is $p(0,y) = p_0$.

1. Denoting by U and V the characteristic orders of magnitude of u and v, respectively, we assume that $V/U = O(\varepsilon)$. Show that $u(x,y)$, $v(x,y)$ and $p(x,y)$ satisfy the lubrication equations (4.11) to (4.14) to first order in ε.

2. Compute $u(x,y)$ in terms of $G(x) = -dp/dx$.

3. Compute $v(x,y)$ and show that

$$G'(x) = dG/dx = -12\mu V_0 f(x)/h^3 \qquad (4.77)$$

4. Find the relation between $Q_x(x)$, $Q_x(0)$ and $V_0 f(x)$, where Q_x is the flow rate per unit width of film (measured along the z-direction). Find again this relation from a mass balance between the entrance section of the film and a section at position x.

5. We now consider a coupling process between the flow in the film and the cross flow through the porous wall (as in industrial processes). The flow across the lower wall is governed by Darcy's law which can be written as

$$v(x,0) = Kp(x)$$

where K is a constant and where the pressure outside the porous wall (somewhere at $y < 0$) is constant and taken to be zero.

6. Find the values of $u(x,y)$ and $v(x,y)$.

7. Give the general expressions of $p(x)$ and $Q_x(x)$. It will be convenient to introduce a characteristic filtration length $L_f = [h^3/(12\mu K)]^{1/2}$.

8. Find $p(x)$ and $Q_x(x)$ when the flow between the two plates is driven by the pressures at the entrance and exit sections $p(0) = p_0$ and $p(L) = 0$ with $p_0 > 0$.

9. Find $p(x)$ and $Q_x(x)$ when the flow between the two plates is driven by the entrance conditions $p(0) = p_0$ and $Q_x(0) = Q_{x0}$. Find the pressure $p(L)$ and the flow rate $Q_x(L)$ at the exit $x = L$.

4.5.4 Motion of a Cylinder between Two Parallel Walls

From an Ecole Polytechnique problem written with K. Moffatt

During extrusion of fibre-reinforced resins, a fibre suspension sometimes flows between two walls separated by a distance which is of the same order of magnitude as the particle characteristic dimension. The following problem addresses this situation in the ideal case of a two-dimensional flow of long fibres between two parallel walls.

Two plates parallel to the xz-plane are located at $y = \pm b$. A solid cylinder with radius a ($a < b$ and $h_0 = b - a \ll a$) and with its axis in the z-direction is placed at the same distance from the two plates. The space between the

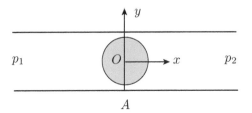

FIGURE 4.17
Cylinder with radius a moving between two parallel plates.

two plates is filled with a Newtonian liquid of viscosity μ (Figure 4.17). The objective of the problem is to study the motion of the cylinder between the two plates. This motion is controlled by the lubrication effects in the film between the cylinder and the walls.

1. Show that in the neighbourhood of the point A in Figure 4.17, the equation of the cylinder surface is approximately given by

$$y + b = h(x) \cong h_0 + x^2/2a$$

2. The cylinder has velocity $U\mathbf{e}_x$ in the fluid which is at rest at infinity $x = \pm\infty$. Show that the flow rate per unit width (along the z-direction) is given by

$$Q_x = -Ua$$

3. Show that the pressure gradient is then

$$\frac{dp}{dx} \cong \frac{12\mu a U}{h^3}$$

4. There is a pressure difference between the far upstream and downstream sections of the channel

$$p \to p_1 \quad \text{for} \quad x \to -\infty, \qquad p \to p_2 \quad \text{for} \quad x \to +\infty$$

   Show that this pressure difference is given by

$$p_2 - p_1 = \frac{9\sqrt{2\pi}\mu U}{2a} \left(\frac{a}{h_0} \right)^{5/2}$$

5. Compute the drag per unit length $F_x\mathbf{e}_x$ exerted by the fluid on the cylinder.

6. The cylinder is now moving in a direction parallel to its axis with velocity $W\mathbf{e}_z$ in a fluid at rest at infinity ($p_1 = p_2 = p_0$). Determine the velocity field $w(x,y)\mathbf{e}_z$ in the fluid and the drag per unit length $F_z\mathbf{e}_z$ on the cylinder.

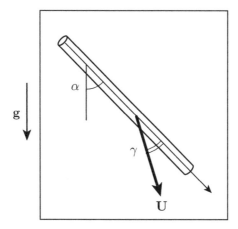

FIGURE 4.18
Free fall of a cylinder between two vertical parallel plates.

7. The two plates are now vertical and the cylinder axis makes an
   angle α with gravity. The cylinder density ρ_s is larger than the
   fluid density, so that the cylinder falls freely between the plates.
   Find the angle γ between the cylinder trajectory and the cylinder
   axis. Discuss the result.

5

Free Surface Films

CONTENTS

We turn now to thin films for which one boundary is a free surface between two immiscible liquids. This situation is commonly encountered in nature (gravity currents, spreading of pollution, eye tear film, etc) or in industry (coating of a surface). The hydrodynamics are then often coupled to heat and mass transport phenomena which modify the physical properties of the flowing fluid. For example, the viscosity of lava increases as it cools down. Similarly, the surface tension and viscosity of a paint film change as the solvent evaporates. Such transport phenomena complicate the analysis of the flow and numerical models are usually necessary.

In this chapter, we consider the flow of thin liquid films with no mass or heat transport. We assume that the condition $\varepsilon Re \ll 1$ is fulfilled, where ε is the ratio between the thickness and the longitudinal dimension of the film.

5.1 Interface between Two Immiscible Fluids: Boundary Conditions

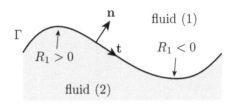

FIGURE 5.1
Free surface Γ between two immiscible fluids.

5.1.1 Geometry of a Free Surface

Two immiscible fluids (1) and (2) are separated by a surface Γ defined by the equation

$$F(x, y, z, t) = 0 \tag{5.1}$$

The unit normal vector $\mathbf{n}$ to Γ points from fluid (2) towards fluid (1) (Figure 5.1). The curvature κ of the interface is defined by

$$\kappa = \nabla \cdot \mathbf{n} \tag{5.2}$$

The curvature can also be defined in terms of the two principal radii of curvature R_1 and R_2 of the surface Γ:

$$\kappa = 1/R_1 + 1/R_2 \tag{5.3}$$

The curvature is either positive or negative as shown in Figure 5.1, but its sign obviously depends on the orientation of the normal vector. The determination of the radii of curvature of a general surface is difficult unless the surface is planar ($\kappa = 0$) or axisymmetric. An axisymmetric surface with revolution axis Oz is defined by the equation of a meridian curve $r = R(z)$ in cylindrical coordinates. The principal radii of curvature are those of the meridian curve

$$R_1 = -(1 + R'^2)^{3/2}/R'' \tag{5.4}$$

and of the parallel circles

$$R_2 = R(1 + R'^2)^{1/2} \tag{5.5}$$

For a sphere, the two radii of curvature are equal to the sphere radius. For a cylinder, one radius is infinite and the other is equal to the cylinder radius.

We note $\mathbf{u}^{(\alpha)}$, $p'^{(\alpha)}$ and $\boldsymbol{\sigma}'^{(\alpha)}$ ($\alpha = 1, 2$), the velocity, hydrodynamic pressure and hydrodynamic stress fields in each fluid.

5.1.2 Kinematic Conditions

The no-slip condition on the free surface leads to the continuity of velocity condition:

$$\mathbf{u}^{(1)} = \mathbf{u}^{(2)} \quad \text{for} \quad \mathbf{x} \in \Gamma \tag{5.6}$$

Furthermore, as the interface is a material surface, the particles on it are convected with the flow, which leads to a kinematic condition on Γ:

$$DF/Dt = 0$$

or

$$\frac{1}{|\nabla F|} \frac{\partial F}{\partial t} + \mathbf{u}^{(\alpha)} \cdot \mathbf{n} = 0 \quad \text{for} \quad \mathbf{x} \in \Gamma \quad \text{and} \quad \alpha = 1, 2 \tag{5.7}$$

where we have used the fact that the unit normal vector is directed along the gradient of F ($\mathbf{n} = \nabla F / |\nabla F|$). Equation (5.7) relates the time evolution of the free surface to the normal velocity of the fluid. When the free surface is stationary, Equation (5.7) simplifies to

$$\mathbf{u}^{(\alpha)} \cdot \mathbf{n} = 0 \quad \text{for} \quad \mathbf{x} \in \Gamma \quad \text{and} \quad \alpha = 1, 2 \tag{5.8}$$

5.1.3 Dynamic Conditions: Laplace's Law

In general, the surface tractions are discontinuous on a liquid-liquid interface. The *hydrodynamic* traction jump across the interface is given by Laplace's law

$$[\boldsymbol{\sigma}'] \cdot \mathbf{n} \equiv \left(\boldsymbol{\sigma}'^{(1)} - \boldsymbol{\sigma}'^{(2)} \right) \cdot \mathbf{n} = \gamma_s \kappa \mathbf{n} \tag{5.9}$$

where the constant surface tension γ_s (N/m) is a physical property of the two-fluid system. For example, $\gamma_s = 72 \times 10^{-3}$ N/m for an air-water interface at 20°C. Equation (5.9) indicates that the normal tractions are discontinuous but that the tangential tractions are continuous across the interface.

It is important to note that Laplace's law applies to the *hydrodynamic* stress jump.

In some circumstances, the surface tension γ_s can vary along the interface. This phenomenon occurs under the influence of a temperature gradient along the interface or when there is a non-homogeneous distribution of adsorbed surfactant molecules on the interface. In general, γ_s decreases when the temperature or the surfactant concentration increases. In this case, the tangential tractions are also discontinuous across the interface and Laplace's law (5.9) is modified to

$$[\boldsymbol{\sigma}'] \cdot \mathbf{n} \equiv \left(\boldsymbol{\sigma}'^{(1)} - \boldsymbol{\sigma}'^{(2)} \right) \cdot \mathbf{n} = \gamma_s \kappa \mathbf{n} - \nabla_s \gamma_s \tag{5.10}$$

where $\nabla_s \gamma_s$ is the gradient of γ_s measured along the interface.

5.1.4 Air–Liquid Interface

When fluid (1) is air, viscous effects are negligible compared to those in the liquid and the stress tensor becomes

$$\boldsymbol{\sigma}'^{(1)} = -p_0 \mathbf{I} \qquad (5.11)$$

where p_0 is the air pressure. In the liquid phase (2), the viscous traction on Γ has a normal component given by

$$\sigma'_{nn} = \sigma'_{ij} n_i n_j = \mathbf{n} \cdot (\boldsymbol{\sigma}' \cdot \mathbf{n}) \qquad (5.12)$$

and a tangential component given by

$$\sigma'_{nt} = \sigma'_{ij} n_j - \sigma'_{nn} n_i = \mathbf{t} \cdot (\boldsymbol{\sigma}' \cdot \mathbf{n}) \qquad (5.13)$$

where $\mathbf{t}$ is a unit vector tangent to Γ (Figure 5.1). In the case of a simple air-liquid interface with no surface tension gradient, Laplace's law (5.9) takes the simple form

$$\sigma'_{nn} = -p_0 - \gamma_s \kappa \quad \text{and} \quad \sigma'_{nt} = 0 \qquad (5.14)$$

5.2 Gravity Spreading of a Fluid on a Horizontal Plane

As an example of free surface moving film, we study the spreading of a fluid volume under the effect of gravity. The objective is to determine the propagation rate. This type of problem is encountered in many practical applications: spreading of lava, of a pollution sheet, of an oil spill, etc. In order to simplify the problem, we consider the axisymmetric spreading of a finite volume of liquid on a horizontal plane. This is a particular case of a whole class of gravity spreading on non-horizontal surfaces with or without continuous feeding of the flow [29].

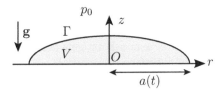

FIGURE 5.2
Axisymmetric gravity spreading of a finite fluid volume V on a horizontal plane.

5.2.1 Spreading with No Surface Tension

A liquid volume V is spreading on a horizontal plane under the effect of gravity. We assume that the liquid domain is axisymmetric about a vertical axis Oz, and we use cylindrical coordinates with axis Oz. Let $h(r,t)$ and $a(t)$ be the thickness and the radius of the film, respectively (Figure 5.2). The equation of the free surface is thus

$$z = h(r,t) \qquad (5.15)$$

We assume that the lubrication conditions are satisfied:

$$h(0,t)/a(t) = \varepsilon \ll 1 \quad \text{and} \quad |\partial h/\partial r| = O(\varepsilon) \qquad (5.16)$$

which implies that to first order, $\mathbf{n} = \mathbf{e}_z + O(\varepsilon)$ on the film surface and that the flow is quasi-bidimensional in planes $z = Cst$

$$\mathbf{u} = u(r,z,t)\mathbf{e}_r \quad \text{and} \quad p' = p'(r,z,t) \qquad (5.17)$$

The atmospheric pressure outside the film is p_0. We neglect surface tension effects. The solution to this problem proceeds in successive steps.

i) Application of the lubrication theory
The problem boundary conditions are

$$u(r,0,t) = 0 \qquad (5.18)$$

and from Equation (5.14)

$$p' = p_0 \quad \text{and} \quad \mu \partial u/\partial z = 0 \quad \text{on} \quad z = h(r,t) \qquad (5.19)$$

The vertical component of the Stokes momentum equation is

$$0 = -\rho g - \partial p'/\partial z \qquad (5.20)$$

which is easily integrated:

$$p' = p_0 + \rho g \left[h(r,t) - z \right] \qquad (5.21)$$

The pressure distribution in the sheet is thus hydrostatic. Using Equation (5.21) we find the radial component of the Stokes momentum equation

$$0 = -\frac{\partial p'}{\partial r} + \mu \frac{\partial^2 u}{\partial z^2} = -\rho g \frac{\partial h}{\partial r} + \mu \frac{\partial^2 u}{\partial z^2} \qquad (5.22)$$

which is also easily integrated:

$$u(r,z,t) = \frac{\rho g}{2\mu} \frac{\partial h}{\partial r} z(z - 2h) \qquad (5.23)$$

It follows that the velocity profile is parabolic with a maximum value on the interface (Figure 5.3). The flow rate in the radial direction is

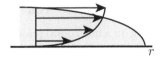

FIGURE 5.3
Parabolic velocity profile in the film.

$$Q(r,t) = 2\pi r \int_0^h u \, dz = -\frac{2\pi\rho g}{3\mu} r \, h^3 \frac{\partial h}{\partial r} \qquad (5.24)$$

The mass conservation Equation (4.21) becomes:

$$\frac{\partial h}{\partial t} = -\frac{1}{2\pi r} \frac{\partial Q}{\partial r} = \frac{g}{3\nu} \frac{1}{r} \frac{\partial}{\partial r} \left(r h^3 \frac{\partial h}{\partial r} \right) \qquad (5.25)$$

with boundary conditions

$$h(a,t) = 0 \quad \text{and} \quad \int_0^{a(t)} h(r,t) \, 2\pi r \, dr = V \qquad (5.26)$$

We thus obtain a non-linear diffusion equation (5.25) for which we seek an asymptotic solution for large time $t \to \infty$. This is equivalent to saying that we neglect the first instants of the flow which depend on how the liquid has been deposited.

ii) Self-similar solution

The classical technique to solve this diffusion problem consists of seeking a *self-similar* solution. We thus introduce two new variables η and $H(\eta)$ defined by

$$\eta = r \, t^{-\beta} \quad \text{and} \quad h(r,t) = t^{-\alpha} H(\eta) \qquad (5.27)$$

It follows that

$$\partial\eta/\partial r = t^{-\beta} \quad \text{and} \quad \partial\eta/\partial t = -\beta\eta/t$$

The evolution equation (5.25) then becomes

$$-(\alpha H + \beta\eta \frac{dH}{d\eta}) t^{-\alpha-1} = \frac{g}{3\nu\eta} t^{-4\alpha-2\beta} \frac{d}{d\eta} \left(\eta H^3 \frac{dH}{d\eta} \right) \qquad (5.28)$$

Since we seek the asymptotic behaviour for $t \to \infty$, we must eliminate time from Equation (5.28). This leads to

$$3\alpha + 2\beta = 1$$

In terms of the new variables, the volume conservation equation (5.26) becomes

$$\int_0^A H(\eta) \, 2\pi\eta \, t^{-\alpha+2\beta} \, d\eta = V \quad \text{with} \quad H(A) = 0 \qquad (5.29)$$

where $A = a/t^\beta$. Eliminating again time leads to

$$-\alpha + 2\beta = 0$$

Thus the self-similar (or time-independent) solution is obtained for

$$\alpha = 1/4 \quad \text{and} \quad \beta = 1/8,$$

Equation (5.25) then becomes a second-order ordinary differential equation for the unknown function $H(\eta)$

$$\frac{d}{d\eta}\left(\eta H^3 \frac{dH}{d\eta}\right) + \frac{3\nu}{8g}\left(2\eta H + \eta^2 \frac{dH}{d\eta}\right) = 0 \tag{5.30}$$

with boundary conditions

$$H(A) = 0 \quad \text{and} \quad H(0) \text{ finite}$$

The solution of Equation (5.30) is

$$H = \left(\frac{9\nu}{16g}\right)^{1/3} (A^2 - \eta^2)^{1/3} \tag{5.31}$$

iii) Result
The volume of the fluid is given by

$$V = \int_0^A 2\pi H \eta \, d\eta$$

Using Equation (5.31) we find

$$V = \frac{3\pi}{4}\left(\frac{9\nu}{16g}\right)^{1/3} A^{8/3} = \frac{3\pi}{4}\left(\frac{9\nu}{16g}\right)^{1/3} a^{8/3} t^{-1/3} \tag{5.32}$$

The radius of the fluid film is then

$$a(t) = 0.779 \left(\frac{t}{T}\right)^{1/8} V^{1/3} \tag{5.33}$$

where $T = \nu/gV^{1/3}$ is a characteristic time.

The film radius varies as $t^{1/8}$ during spreading. This spreading law has been verified experimentally [29]. The film profile, shown in Figure 5.4, is determined by Equations (5.31) and (5.32). The tangent to the profile is vertical at $r = a$, which is not realistic and does not correspond to experimental observations. As already mentioned in Chapter 4, the lubrication hypotheses are no longer satisfied at a distance $O(\varepsilon a)$ from the contact line between the film and the plane. In particular, the hypothesis $|u| \gg |w|$ is not satisfied there. The self-similar solution is thus valid only for films with a large radius-to-thickness ratio. In this case, the details of the flow near the contact line are of limited importance.

It is possible to generalise this axisymmetric analysis and consider instead the spreading of a film on a conical surface with a vertical axis (or any other axisymmetric surface). We just have to replace **g** by its projection on the surface. We then can model (very approximately!) the lava flow from an erupting volcano or the more prosaic icing of a cake.

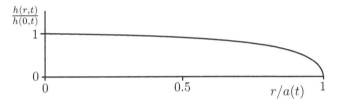

FIGURE 5.4
Film profile during gravity spreading of a finite fluid volume on a horizontal plane.

5.2.2   Effect of Surface Tension

When surface tension is not negligible, the boundary condition (5.19) must be modified

$$p' = p_0 + \gamma_s \kappa \quad \text{and} \quad \mu \partial u / \partial z = 0 \quad \text{on} \quad z = h(r,t) \tag{5.34}$$

The pressure in the film becomes

$$p' = p_0 + \rho g \left[h(r,t) - z \right] + \gamma_s \kappa$$

where the surface curvature is given to first order in ε by

$$\kappa \cong -\frac{\partial^2 h}{\partial r^2} - \frac{1}{r}\frac{\partial h}{\partial r}$$

The time evolution equation of the film thickness is then

$$\frac{\partial h}{\partial t} = \frac{g}{3\nu}\frac{1}{r}\frac{\partial}{\partial r}\left[rh^3 \left(\frac{\partial h}{\partial r} + \frac{\gamma_s}{\rho g}\frac{\partial \kappa}{\partial r} \right) \right] \tag{5.35}$$

There is no self-similar solution of this equation which must be solved numerically. It is possible to compare the order of magnitude of the two terms on the right-hand side of Equation (5.35)

$$\left| \frac{\partial h}{\partial r} \right| \sim \frac{h}{a} \quad \text{and} \quad \left| \frac{\gamma_s}{\rho g}\frac{\partial \kappa}{\partial r} \right| \sim \frac{\gamma_s h}{\rho g a^3}$$

It follows that

$$\left|\frac{\partial h}{\partial r}\right| \Big/ \left|\frac{\gamma_s}{\rho g}\frac{\partial \kappa}{\partial r}\right| \sim \frac{\rho g a^2}{\gamma_s}$$

The ratio between the gravity and surface tension terms is measured by the Bond number B

$$B = \rho g a^2/\gamma_s = a^2/l_c^2$$

where the capillary length l_c is a physical property of the air-liquid system

$$l_c^2 = \gamma_s/\rho g$$

For an air-water system, the capillary length is 'large', of order 2.6 mm. We can then conclude that during the spreading of a liquid film with a radius much larger than l_c, the surface tension effects are negligible. In contrast, for small liquid volumes of order l_c^3, the surface tension and the contact angle between the interface and the solid substrate become the principal effects that govern the final form of the drop. But of course we are then out of the realm of lubrication theory.

5.3 Stability of a Film Flowing Down an Inclined Plane

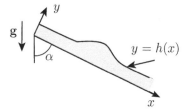

FIGURE 5.5
Gravity flow of a film down an inclined plane.

Another example of free surface flows is frequently encountered when a liquid film flows on an inclined surface. This situation occurs after the paint film deposited by a scraper (or a brush) is allowed to flow under gravity (see Chapter 4, Section 3.3) or when a liquid film is used to cool off a surface.

A Newtonian incompressible liquid (viscosity μ and density ρ) flows along a flat plate making an angle α ($\alpha \geq 0$) with the vertical direction (Figure 5.5). The atmospheric pressure p_0 outside the film is uniform and surface tension effects are neglected. The flow is assumed to be two-dimensional in the xy-plane. The film surface is defined by

$$y = h(x,t)$$

where the film thickness $h(x,t)$ may vary in space and time. We assume that the lubrication conditions are satisfied,

$$|h|/L \ll 1 \qquad \text{and} \qquad |\partial h/\partial x| \ll 1$$

where L is the characteristic film length in the x-direction. The flow is thus quasi-unidirectional

$$\mathbf{u} = u(x,y,t)\mathbf{e}_x \qquad \text{and} \qquad p = p(x,y,t)$$

When there is no surface tension, the boundary conditions are similar to those of Section 5.2

$$u(x,0,t) = 0 \tag{5.36}$$

$$p' = p_0 \quad \text{and} \quad \mu \partial u/\partial y = 0 \quad \text{on} \quad y = h(x,t)$$

The equation of motion in the y-direction leads to

$$0 = -\rho g \sin \alpha - \partial p'/\partial y \tag{5.37}$$

which can be integrated

$$p' = p_0 + \rho g \sin \alpha (h - y) \tag{5.38}$$

The equation of motion in the x-direction leads to

$$0 = -\frac{\partial p'}{\partial x} + \rho g \cos \alpha + \mu \frac{\partial^2 u}{\partial y^2} = -\rho g \sin \alpha \frac{\partial h}{\partial x} + \rho g \cos \alpha + \mu \frac{\partial^2 u}{\partial y^2} \tag{5.39}$$

where we have used Equation (5.38). Integrating Equation (5.39) allows us to determine the velocity field in the film:

$$u = -\frac{g}{2\nu} \left(\cos \alpha - \sin \alpha \frac{\partial h}{\partial x} \right) y(y - 2h) \tag{5.40}$$

and to deduce the flow rate Q per unit width of film

$$Q = \int_0^h u \, dy = \frac{g}{3\nu} \left(\cos \alpha - \sin \alpha \frac{\partial h}{\partial x} \right) h^3 \tag{5.41}$$

When the film thickness is constant along the x-direction, the thickness h_0 is determined by the feed flow rate Q

$$Q = \frac{g \cos \alpha}{3\nu} h_0^3 \tag{5.42}$$

This control relation allows us to determine the flow rate necessary to obtain a given film thickness as a function of plate angle and fluid viscosity.

When the film thickness varies in space and time, mass conservation leads to a time evolution equation for the thickness $h(x,t)$:

$$\frac{\partial h}{\partial t} = \frac{g}{3\nu}\frac{\partial}{\partial x}\left[h^3\left(\sin\alpha\frac{\partial h}{\partial x} - \cos\alpha\right)\right] \qquad (5.43)$$

This non-linear equation must be solved numerically in general. However, we can consider the time and space evolution of a small perturbation $\Delta h(x,t)$ ($|\Delta h| \ll h_0$) from the constant thickness film:

$$h(x,t) = h_0 + \Delta h(x,t) \qquad (5.44)$$

The time evolution equation of the perturbation becomes

$$\frac{\partial \Delta h}{\partial t} = \frac{g h_0^3}{3\nu}\sin\alpha\frac{\partial^2 \Delta h}{\partial x^2} - \frac{g h_0^2}{\nu}\cos\alpha\frac{\partial \Delta h}{\partial x} \qquad (5.45)$$

The first term on the right-hand side of Equation (5.45) is a diffusion term that tends to flatten the perturbation $\Delta h(x,t)$. The second term on the right-hand side tends to increase the perturbation $\Delta h(x,t)$. It follows that for $\alpha = \pi/2$ (horizontal plane) the film is stable, but that for $\alpha = 0$ (vertical plane) the film is always unstable. This instability is well known: when you paint a vertical surface and deposit too much paint, the film is thick, dries slowly and this allows the instability to develop as unseemly dripping.

A complicated detailed analysis of the film stability allows us to state that the film is stable if [10]

$$h_0 < \left(\frac{10\nu^2}{4g}\frac{\tan\alpha}{\cos\alpha}\right)^{1/3} \qquad (5.46)$$

In the case of a film flowing under a plane $\alpha \leq 0$, the h evolution equation (5.43) is always unstable and mathematically ill-posed. It can nevertheless model the first instants of the dripping instability under a plate. Of course, we all know that it is possible to paint surfaces under reverse gravity (e.g. ceilings, three-dimensional complicated objects). In that case, the film must be thin so that the solvent evaporates fast enough to prevent the dripping instability from appreaing (note that the viscosity increases with evaporation and the stability criterion (5.46) increases accordingly).

Finally, when we take into account surface tension effects, the problem becomes even more complicated. However, we can expect a stabilising influence of capillary forces. Indeed, since such forces oppose an increase of the film surface, they will thus oppose the growth of a thickness instability.

5.4   Problems

5.4.1   Gravity Spreading of a Film with Continuous Flux

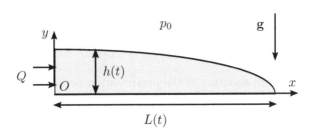

FIGURE 5.6
Two-dimensional gravity flow of a thin layer with continuous feed flux Q.

We consider the two-dimensional gravity flow of a liquid layer of a heavy liquid which is continuously fed by a flux Q. This problem is an approximate model of the flow of cold air under a door, the flow of muddy effluents at the bottom of a reservoir or the progression of an oil spill. The liquid layer flows along the horizontal x-direction and is 'infinite' in the other horizontal perpendicular z-direction (Figure 5.6). The velocity field is quasi-unidirectional, $\mathbf{u} = u(x, y, t)\mathbf{e}_x$. The film thickness, measured along the vertical y-direction, is $h(x, t)$. The surrounding fluid is air with negligible density and viscosity compared to those of the liquid and with uniform pressure p_0. The layer is supplied with liquid at $x = 0$ with a constant flow rate Q per unit width (along Oz). We denote $L(t)$ the length of the layer in the x-direction and we assume that $h/L \ll 1$. Surface tension effects are ignored.

1. Write the equation of motion and the boundary conditions.

2. Find the time and space evolution equation for $h(x, t)$.

3. Express the total conservation of volume at time t.

4. Seek a self-similar solution to the equations and show that

$$L \propto t^{4/5}$$

Remark: The resulting equation has no analytical solution; do not seek one!

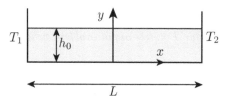

FIGURE 5.7
Thermocapillary flow in a 2D cavity under the influence of a temperature difference $T_2 < T_1$.

5.4.2 Two-Dimensional Thermocapillary Flow

Capillary forces due to a gradient of surface tension can exert a large enough traction on the surface of a liquid film to create a bulk flow inside the film. Such effects (called Marangoni effects) are found either when a temperature or a concentration gradient is created on the interface. The objective of this problem is to study such an effect in a simple two-dimensional configuration.

A two-dimensional shallow cavity with rectangular cross section is filled with a liquid with viscosity μ and thermal conductivity λ that are both much larger than those of air (pressure p_0). The liquid depth h_0 is small with respect to the cavity width L (Figure 5.7). The lateral wall $x = -L/2$ is at temperature T_1 and the other wall $x = L/2$ is at temperature T_2 ($T_2 < T_1$). The bottom of the cavity is thermally isolated. The surface tension γ_s between the liquid and air varies linearly with temperature,

$$\gamma_s = \gamma_{s0} - \beta(T - \frac{T_1 - T_2}{2})$$

where γ_{s0} is the surface tension at temperature $T = (T_1 - T_2)/2$. There is thus an interfacial surface tension gradient which exerts a traction on the fluid and sets it into motion. We neglect gravity effects and seek the flow in the cavity as well as the shape of the interface.

1. Find the surface tension on the interface, assuming that the heat transport is conductive.

2. Write the equations of motion and the boundary conditions.

3. Compute the flow field in the central part of the cavity (i.e. far from the walls).

4. Sketch the flow pattern in the cavity.

5. Show that the interface cannot remain horizontal and find its equation.

5.4.3 Motion of a Contact Lens on the Eye

From an Ecole Polytechnique problem written with J. Magnaudet

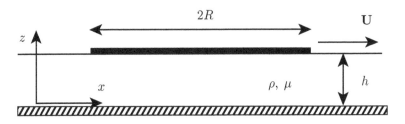

FIGURE 5.8
Motion of a solid disk on a thin liquid film.

The following problem is a simplified model of the motion of a contact lens on the eye.

A solid wall at $z = 0$ is covered by a thin film (thickness h) of an incompressible Newtonian liquid (viscosity μ, density ρ). In the lateral x- and y-directions, the film dimensions are order L, such that $L \gg h$. A solid disk of radius R ($R \gg h$), with negligible thickness and mass, lies on the film surface at $z = h$. This disk is put into motion with a constant velocity $\mathbf{U} = U\mathbf{e}_x$ (Figure 5.8). The objective of the problem is to determine the velocity $\mathbf{u}(x, y, z)$ and pressure $p(x, y, z)$ in the liquid, as well as the force that must be exerted on the disk.

We use the lubrication theory and we neglect gravity forces. We neglect the deformation of the film interface and the liquid velocity in the z-direction. We first study the general flow in a liquid film when the surface velocity is imposed. We then use the results to analyse the flow under the disk and in the outer film.

1. Flow in a uniform liquid film with imposed surface velocity

    We consider the flow in a liquid film of uniform thickness h under the effect of a given surface velocity $\mathbf{u}_s(x, y)$ such that

    $$\mathbf{u}(x, y, h) = \mathbf{u}_s(x, y)$$

    (a) Write the equations of motion and the boundary conditions for the flow in the film.

    (b) Find the velocity $\mathbf{u}$ and the flow rate $\mathbf{Q} = \int_0^h \mathbf{u}\, dz$ in terms of $\mathbf{u}_s$ and of the 2D pressure gradient $\nabla' p = \partial p/\partial x\, \mathbf{e}_x + \partial p/\partial y\, \mathbf{e}_y$.

    (c) Use mass conservation to deduce that

    $$\nabla'^2 p = \frac{6\mu}{h^2} \nabla' \cdot \mathbf{u}_s$$

(d) The traction on the film surface is defined as $\mathbf{T} = \mu \left(\partial \mathbf{u}/\partial z \right)_{z=h}$. Find the general expression for $\mathbf{T}$ in terms of $\mathbf{u}_s$ and ∇p.

We now use a polar coordinate system (r, θ) centred on the disk as shown in Figure 5.9 and use the results obtained in Question 1.

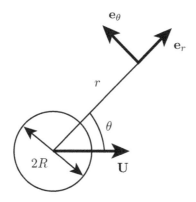

FIGURE 5.9
Polar coordinates centred on the disk.

2. Flow under the disk

   (a) Write the boundary conditions for the flow in the film under the disk $(r < R)$.
   (b) Show that $\nabla^2 p = 0$.
   (c) We seek a solution of the form $p(r, \theta) = U g_{in}(r) \cos \theta$. Find $g_{in}(r)$. Denote A the integration constant that appears in $g_{in}(r)$.

3. Flow outside the disk

   (a) Write the boundary conditions for the flow in the free surface film outside the disk $(r > R)$.
   (b) Show that $\nabla^2 p = 0$.
   (c) We seek again a solution of the form $p(r, \theta) = U g_{ext}(r) \cos \theta$. Find $g_{ext}(r)$. Denote B the integration constant that appears in $g_{ext}(r)$.

4. Show that $B = AR^2$.

5. Find the components of the flow rate $\mathbf{Q}$ in the film in terms of A and $\mathbf{U}$.

6. In reality, the liquid film is bounded by lateral walls which are normal to the direction of motion of the disk. The distance between these walls is much larger than R. Write the condition due to the

presence of walls in the median plane ($\theta = \pi/2$). Deduce the value of A.

7. Find the drag force exerted by the fluid on the disk (assume that the effect of the lateral walls is negligible).

8. Compute the components of $\mathbf{Q}$ in the local r- and θ-directions. Show that only one component is continuous for $r = R$. What is the origin of the discontinuity? How should we proceed to remove it?

9. This problem is a simple model of a contact lens on the eye. Discuss the validity of the hypotheses and of the analysis within the context of this application.

6

Motion of a Solid Particle in a Fluid

CONTENTS

This chapter deals with the motion of a solid particle in a liquid. We assume that the characteristic dimension of the particle is small, so that the Reynolds number of the flow around it is small too. This situation is encountered in many industrial or natural situations involving flowing particle suspensions for which different problems arise: determination of the effective transport properties, stability and sedimentation. We first establish general relations between the particle velocity and the hydrodynamic resultant force which is exerted (whether it is a propulsion or a drag force). These relations are then specified for different particle shapes such as spheres, rods and helices. In a final part we show how the approach presented in this chapter can be used to study the propulsion of micro-organisms with applications in biology, physiology or the environment.

6.1   Motion of a Solid Particle in a Quiescent Fluid

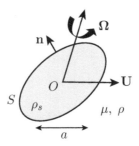

FIGURE 6.1
Solid particle moving in a fluid at rest.

The objective of this section is to use the properties of Stokes flows (Chapter 2) to find some general features of the motion of a particle in a quiescent fluid. A solid particle with density ρ_s is suspended in an incompressible viscous liquid with density ρ and viscosity μ (Figure 6.1). The particle has a characteristic dimension denoted a that can be chosen as the radius of the sphere which has the same volume V as the particle

$$a = (3V/4\pi)^{1/3}$$

In some cases, other choices can be more relevant, specifically for particles that are very anisotropic (see Section 6.5). The particle surface is denoted S with outer unit normal vector denoted $\mathbf{n}$. We use a reference frame linked to the fluid at infinity but centred on the particle centre of mass O at time t

$$\int_V \rho_s \mathbf{x}\, dV = 0 \tag{6.1}$$

If the particle is homogeneous, O is its geometrical centre. Far from the particle, the fluid is at rest. The particle has a solid body motion with translation velocity $\mathbf{U}$ and angular velocity $\mathbf{\Omega}$:

$$\mathbf{u}^{(S)} = \mathbf{U} + \mathbf{\Omega} \times \mathbf{x}$$

where $\mathbf{u}^{(S)}$ is the velocity of a point of the particle located at $\mathbf{x}$. This motion generates a surrounding flow with a very small Reynolds number because the particle is very small

$$Re = \frac{\rho |\mathbf{U}| a}{\mu} \ll 1 \quad \text{and} \quad Re = \frac{\rho |\mathbf{\Omega}| a^2}{\mu} \ll 1 \tag{6.2}$$

The flow of the fluid around the particle thus satisfies the Stokes equations

$$\nabla \cdot \mathbf{u} = 0 \quad \text{and} \quad \nabla p = \mu \nabla^2 \mathbf{u} \tag{6.3}$$

with boundary conditions

$$\mathbf{u} \to 0 \quad \text{for} \quad |\mathbf{x}| \to \infty \tag{6.4}$$

$$\mathbf{u} = \mathbf{U} + \boldsymbol{\Omega} \times \mathbf{x} \quad \text{for} \quad \mathbf{x} \in S \tag{6.5}$$

This Stokes problem has an analytical solution for simple geometries such as spheres, ellipsoids, slender particles, etc. In general, a numerical solution must be sought using one of the different techniques presented in Chapter 8. However, the linearity of the Stokes equations allows us to infer some general properties of the solution.

6.1.1 Resistance and Mobility Tensors

The particle is subjected to a resultant force $\mathbf{F}$ and torque $\mathbf{G}$ created by the hydrodynamic stresses due to the fluid motion:

$$\mathbf{F} = \int_S \boldsymbol{\sigma} \cdot \mathbf{n} \, dS \quad \text{or} \quad F_i = \int_S \sigma_{ij} n_j \, dS \tag{6.6}$$

$$\mathbf{G} = \int_S \mathbf{x} \times (\boldsymbol{\sigma} \cdot \mathbf{n}) \, dS \quad \text{or} \quad G_i = \int_S \varepsilon_{ijk} x_j (\sigma_{km} n_m) \, dS \tag{6.7}$$

Since the Stokes equations are linear, the velocity $\mathbf{u}$ and the stress $\boldsymbol{\sigma}$ in the fluid are linear functions of the causes of motion $\mathbf{U}$ and $\boldsymbol{\Omega}$. It follows that the resultant force $\mathbf{F}$ and torque $\mathbf{G}$ are also linear functions of $\mathbf{U}$ and $\boldsymbol{\Omega}$. The following general linear relationships can then be written:

$$\mathbf{F} = -\mu(a\mathbf{A} \cdot \mathbf{U} + a^2 \mathbf{B} \cdot \boldsymbol{\Omega}) \quad \text{or} \quad F_i = -\mu \left(a \, A_{ij} U_j + a^2 B_{ij} \Omega_j \right) \tag{6.8}$$

$$\mathbf{G} = -\mu(a^2 \mathbf{C} \cdot \mathbf{U} + a^3 \mathbf{D} \cdot \boldsymbol{\Omega}) \quad \text{or} \quad G_i = -\mu \left(a^2 \, C_{ij} U_j + a^3 D_{ij} \Omega_j \right) \tag{6.9}$$

where $\mathbf{A}$, $\mathbf{B}$, $\mathbf{C}$ and $\mathbf{D}$ are called the resistance tensors. The minus sign indicates that the fluid resists the motion of the particle. It is possible to invert relations (6.8) and (6.9) to relate the particle velocity to the resultant force and torque exerted on it:

$$\mathbf{U} = -\frac{1}{\mu a^2} \left(a\mathbf{A}' \cdot \mathbf{F} + \mathbf{B}' \cdot \mathbf{G} \right) \quad \text{or} \quad U_i = -\frac{1}{\mu a^2} \left(a \, A'_{ij} F_j + B'_{ij} G_j \right) \tag{6.10}$$

$$\boldsymbol{\Omega} = -\frac{1}{\mu a^3} \left(a\mathbf{C}' \cdot \mathbf{F} + \mathbf{D}' \cdot \mathbf{G} \right) \quad \text{or} \quad \Omega_i = -\frac{1}{\mu a^3} \left(a \, C'_{ij} F_j + D'_{ij} G_j \right) \tag{6.11}$$

where $\mathbf{A}'$, $\mathbf{B}'$, $\mathbf{C}'$ and $\mathbf{D}'$ are called the mobility tensors.

The resistance tensors $\mathbf{A}$, $\mathbf{B}$, $\mathbf{C}$ and $\mathbf{D}$ or the mobility tensors $\mathbf{A}'$, $\mathbf{B}'$, $\mathbf{C}'$ and $\mathbf{D}'$ are dimensionless and depend only on the particle geometry. They are computed (analytically or numerically) only once for a particle with a given shape.

The particle motion is necessarily due to an external cause (gravity, electric force on a charged particle, magnetic force, etc) which exerts a resultant force $\mathbf{F}_e$ and a torque $\mathbf{G}_e$. If the particle inertia can be neglected (implying $\rho_s Re/\rho \ll 1$), the equation of motion becomes

$$\mathbf{F} + \mathbf{F}_e = 0 \quad \text{and} \quad \mathbf{G} + \mathbf{G}_e = 0 \tag{6.12}$$

Consequently, if we know the mobility tensors of the particle and the external forces to which it is subjected, the relations (6.10), (6.11) and (6.12) allow us to compute its velocity $(\mathbf{U}, \mathbf{\Omega})$:

$$\mathbf{U} = +\frac{1}{\mu a^2}\left(a\mathbf{A}' \cdot \mathbf{F}_e + \mathbf{B}' \cdot \mathbf{G}_e\right) \tag{6.13}$$

$$\mathbf{\Omega} = +\frac{1}{\mu a^3}\left(a\mathbf{C}' \cdot \mathbf{F}_e + \mathbf{D}' \cdot \mathbf{G}_e\right) \tag{6.14}$$

Conversely, relations (6.8) and (6.9) allow us to compute the external force and torque that must be applied to obtain a given motion of the particle.

6.1.2 Relations between the Resistance and Mobility Tensors

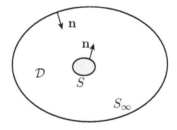

FIGURE 6.2
The fluid domain $\mathcal{D}$ is bounded by the particle surface S and by a surface S_∞ located infinitely far from the particle.

The resistance tensors have symmetry properties that follow from the properties of the Stokes equations (Chapter 2). To demonstrate them, we use the reciprocal theorem for two different flows (1) and (2) of the same fluid in the same domain $\mathcal{D}$ with boundary $\partial\mathcal{D}$:

$$\int_{\partial\mathcal{D}} \left[u_i^{(1)}\sigma_{ij}^{(2)}n_j\right] dS = \int_{\partial\mathcal{D}} \left[u_i^{(2)}\sigma_{ij}^{(1)}n_j\right] dS \tag{6.15}$$

The domain $\mathcal{D}$ is bounded by the particle surface S and by a surface S_∞

located at an infinite distance from the particle (Figure 6.2). The boundary $\partial \mathcal{D}$ is thus $S \cup S_\infty$. On S_∞, $dS \to O(r^2)$ and the perturbation due to the particle vanishes as $|\mathbf{u}| \to O(1/r)$ and $|\boldsymbol{\sigma}| \to O(1/r^2)$, where $r = |\mathbf{x}|$ is the distance between the particle centre and S_∞ (Figure 6.2). Thus the integrals on S_∞ vanish as $r \to \infty$:

$$\int_{S_\infty} \left[u_i^{(1)} \sigma_{ij}^{(2)} n_j \right] dS = 0 \quad \text{and} \quad \int_{S_\infty} \left[u_i^{(2)} \sigma_{ij}^{(1)} n_j \right] dS = 0 \tag{6.16}$$

The reciprocal theorem then becomes here

$$\int_S \left[u_i^{(1)} \sigma_{ij}^{(2)} n_j \right] dS = \int_S \left[u_i^{(2)} \sigma_{ij}^{(1)} n_j \right] dS \tag{6.17}$$

We now consider two translation motions of the particle with velocities $\mathbf{U}_1$ and $\mathbf{U}_2$. The resulting flow fields (1) and (2) satisfy Equation (6.3) with boundary conditions

$$\mathbf{u}^{(\alpha)} \to 0 \quad \text{for} \quad |\mathbf{x}| \to \infty, \quad \alpha = 1, 2 \tag{6.18}$$

$$\mathbf{u}^{(1)}(\mathbf{x}) = \mathbf{U}^{(1)} \quad \text{and} \quad \mathbf{u}^{(2)}(\mathbf{x}) = \mathbf{U}^{(2)} \quad \text{for} \quad \mathbf{x} \in S \tag{6.19}$$

The reciprocal theorem in Equation (6.17) becomes

$$U_i^{(1)} \int_S \sigma_{ij}^{(2)} n_j \, dS = U_i^{(2)} \int_S \sigma_{ij}^{(1)} n_j \, dS \tag{6.20}$$

or

$$U_i^{(1)} F_i^{(2)} = U_i^{(2)} F_i^{(1)} \tag{6.21}$$

where $F^{(\alpha)}$ is the force exerted on the particle by flow (α). Replacing $F^{(\alpha)}$ with expression (6.8), we find

$$-\mu a U_i^{(1)} A_{ik} U_k^{(2)} = -\mu a U_i^{(2)} A_{ik} U_k^{(1)} \tag{6.22}$$

It follows that the tensor A_{ij} is symmetric:

$$A_{ij} = A_{ji} \quad \text{or} \quad \mathbf{A} = {}^\top\mathbf{A} \tag{6.23}$$

Similarly, with flow fields (1) and (2) defined by

$$\mathbf{u}^{(1)} = \mathbf{U}^{(1)} \quad \text{and} \quad \mathbf{u}^{(2)} = \boldsymbol{\Omega}^{(2)} \times \mathbf{x} \quad \text{for} \quad \mathbf{x} \in S$$

we obtain from Equation (6.17)

$$U_i^{(1)} \int_S \sigma_{ij}^{(2)} n_j \, dS = \Omega_j^{(2)} \int_S \varepsilon_{ijk} x_k \sigma_{im}^{(1)} n_m \, dS = \Omega_j^{(2)} \int_S \varepsilon_{jki} x_k (\sigma_{im}^{(1)} n_m) \, dS$$

$$U_i^{(1)} F_i^{(2)} = \Omega_j^{(2)} G_j^{(1)} = \Omega_i^{(2)} G_i^{(1)}$$

It follows from Equations (6.8) and (6.9) that $\mathbf{B}$ is equal to the transpose of $\mathbf{C}$

$$B_{ij} = C_{ji} \quad \text{or} \quad \mathbf{B} = {}^{\top}\mathbf{C} \tag{6.24}$$

The same demonstration with

$$\mathbf{u}^{(1)} = \mathbf{\Omega}^{(1)} \times \mathbf{x} \quad \text{and} \quad \mathbf{u}^{(2)} = \mathbf{\Omega}^{(2)} \times \mathbf{x} \quad \text{for} \quad \mathbf{x} \in S$$

shows that $\mathbf{D}$ is symmetric

$$D_{ij} = D_{ji} \quad \text{or} \quad \mathbf{D} = {}^{\top}\mathbf{D} \tag{6.25}$$

In conclusion, the motion of a solid body in a fluid is completely defined by 21 $(6 + 9 + 6)$ parameters.

6.1.3   Translation without Rotation

We consider the case where at time t, the particle has only a translation velocity with no rotation:

$$\mathbf{u} = \mathbf{U} \quad \text{and} \quad \mathbf{\Omega} = 0 \quad \text{for} \quad \mathbf{x} \in S$$

The force exerted by the fluid on the particle is given by

$$\mathbf{F} = -\mu a \, \mathbf{A} \cdot \mathbf{U} \quad \text{or} \quad F_i = -\mu a \, A_{ij} \, U_j \tag{6.26}$$

However, a torque $\mathbf{G} = -\mu a^2 \mathbf{C} \cdot \mathbf{U}$ is also exerted by the fluid and tends to rotate the particle (this phenomenon is observed for maple seeds that rotate while they fall, although this is not a low Reynolds number flow). In order to prevent the rotation, it is necessary to exert an external torque $\mathbf{G}_e = +\mu a^2 \mathbf{C} \cdot \mathbf{U}$ which is equal and opposite to the viscous one.

Since the tensor $\mathbf{A}$ is symmetric, there is a reference frame (linked to the particle) in which $\mathbf{A}$ is diagonal

$$\mathbf{A} = \begin{pmatrix} \lambda_1 & 0 & 0 \\ 0 & \lambda_2 & 0 \\ 0 & 0 & \lambda_3 \end{pmatrix} \tag{6.27}$$

The energy dissipation in the fluid at time t is given by

$$\Phi = \mathbf{F}_e \cdot \mathbf{U} = -\mathbf{F} \cdot \mathbf{U} = \mu a \, A_{ij} U_j U_i \tag{6.28}$$

As this dissipation must be non-negative:

$$\Phi = \mu a \left(\lambda_1 U_1^2 + \lambda_2 U_2^2 + \lambda_3 U_3^2 \right) \geqslant 0 \tag{6.29}$$

all the eigenvalues of $\mathbf{A}$ are non-negative:

$$\lambda_1 \geqslant 0, \quad \lambda_2 \geqslant 0, \quad \lambda_3 \geqslant 0 \tag{6.30}$$

6.1.4 Rotation without Translation

We consider the case where at time t, the particle has only a rotation motion with no translation

$$\mathbf{u} = \mathbf{\Omega} \times \mathbf{x} \quad \text{and} \quad \mathbf{U} = 0 \quad \text{for} \quad \mathbf{x} \in S$$

The torque exerted by the fluid on the particle is

$$\mathbf{G} = -\mu a^3 \mathbf{D} \cdot \mathbf{\Omega} \quad \text{or} \quad G_i = -\mu a^3 D_{ij} \Omega_j$$

However, the fluid also exerts a hydrodynamic force $\mathbf{F} = -\mu a^2 \mathbf{B} \cdot \mathbf{\Omega}$ on the particle. In order to prevent the translation of the particle, it is necessary to apply a force $\mathbf{F}_e = +\mu a^2 \mathbf{B} \cdot \mathbf{\Omega}$ which is equal and opposite to the hydrodynamic one.

Since the tensor $\mathbf{D}$ is symmetric, there is a reference frame (linked to the particle) in which $\mathbf{D}$ is diagonal

$$\mathbf{D} = \begin{pmatrix} \eta_1 & 0 & 0 \\ 0 & \eta_2 & 0 \\ 0 & 0 & \eta_3 \end{pmatrix} \tag{6.31}$$

The energy dissipation at time t is given by

$$\Phi = \mathbf{G}_e \cdot \mathbf{\Omega} = -\mathbf{G} \cdot \mathbf{\Omega} = \mu a^3 \, D_{ij} \Omega_j \Omega_i \tag{6.32}$$

As $\Phi \geqslant 0$, the eigenvalues of $\mathbf{D}$ are also non-negative:

$$\eta_1 \geqslant 0, \quad \eta_2 \geqslant 0, \quad \eta_3 \geqslant 0 \tag{6.33}$$

6.2 Isotropic Particles

We now apply the general relation obtained in the preceding section to the special case of isotropic particles. We first consider the motion of a rigid sphere with radius a. In view of the particle isotropy, the eigenvalues of $\mathbf{A}$ and $\mathbf{D}$ are all equal:

$$\lambda_1 = \lambda_2 = \lambda_3 = \lambda \quad \text{and} \quad \eta_1 = \eta_2 = \eta_3 = \eta$$

Furthermore, the reversibility of Stokes flows allows us to prove easily that for isotropic particles $\mathbf{B} = 0$ and $\mathbf{C} = 0$ (see Problem 6.6.1). The resistance tensors then become

$$\mathbf{A} = \lambda \mathbf{I} \quad \text{or} \quad A_{ij} = \lambda \, \delta_{ij}$$
$$\mathbf{D} = \eta \mathbf{I} \quad \text{or} \quad D_{ij} = \eta \, \delta_{ij} \tag{6.34}$$
$$\mathbf{B} = {}^{\mathsf{T}}\mathbf{C} = 0$$

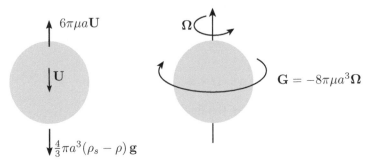

FIGURE 6.3
Sphere undergoing translation or rotation in a fluid at rest.

A sphere translating in a fluid is thus subjected to a resultant viscous force
and torque given by
$$\mathbf{F} = -\lambda\mu a \mathbf{U} \quad \text{and} \quad \mathbf{G} = 0 \qquad (6.35)$$
There is no torque acting on the sphere. Consequently the sphere will not
rotate as an anisotropic particle would (Figure 6.3).

The above reasoning is based on general considerations (reciprocity,
isotropy, reversibility, dimensional analysis), but does not allow us to compute the value of λ. In order to determine the exact value of λ, it is necessary
to compute the flow about a sphere translating in a fluid at rest, a non-trivial
problem (see Section 6.3). Anticipating the result $\lambda = 6\pi$, we obtain the famous Stokes formula which gives the drag force on a sphere translating with
velocity $\mathbf{U}$:
$$\mathbf{F} = -6\pi\mu a \mathbf{U} \qquad (6.36)$$
When a homogeneous sphere is settling under the action of gravity, the external force is
$$\mathbf{F}_e = \frac{4}{3}\pi a^3 (\rho_s - \rho)\mathbf{g}$$
The sedimentation velocity of the sphere follows from Equations (6.12) and
(6.36):
$$\mathbf{U} = \frac{2}{9}\frac{a^2}{\mu}(\rho_s - \rho)\mathbf{g} \qquad (6.37)$$
We find that the sedimentation velocity of a sphere is proportional to the
square of its radius (not an intuitive guess). This sedimentation law is well
verified experimentally. Of course, if the sphere were not homogeneous, it
would take a rotational motion when sedimenting.

Similarly, a rotating sphere is subjected to a resistant hydrodynamic torque
which is co-linear to the rotation vector (Figure 6.3):
$$\mathbf{G} = -\eta\mu a^3 \mathbf{\Omega} \quad \text{and} \quad \mathbf{F} = 0 \qquad (6.38)$$

The value of the resistance coefficient $\eta = 8\pi$ follows from the computation of the flow field around the sphere (see Section 6.4).

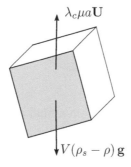

$$\lambda_c \mu a \mathbf{U}$$

$$V(\rho_s - \rho)\,\mathbf{g}$$

FIGURE 6.4
A homogeneous cube with side $2a$ falls under gravity without rotating.

A homogenous cube $2a \times 2a \times 2a$ is also isotropic with respect to its three axes of inertia, centred on the cube centre. We thus obtain the same results as for a sphere $\mathbf{B} = \mathbf{C} = 0$. Consequently, a homogenous cube falls without rotating, irrespective of its initial orientation (Figure 6.4). A reasoning based on the energy dissipation in a reservoir (see Problem 2.8.3, Chapter 2) allows us to estimate the value of the coefficient λ_c for a cube. The energy dissipation due to a falling cube is somewhere between the dissipations due to the inscribed or circumscribed spheres (radius a and $a\sqrt{3}$, respectively) falling with the same velocity U as the cube:

$$1 < \lambda_c/6\pi < \sqrt{3}$$

The determination of the exact value of λ_c must be done numerically, but is difficult to get because of the presence of singularities at the corners.

Remark: Other particles such as polyhedrons can also be isotropic with respect to the three axes of inertia.

6.3 Flow around a Translating Sphere

As noted earlier, the only way to determine the resistance coefficient λ for a translating sphere is to compute the flow field around it. This computation is now presented in detail.

A sphere with radius a is translating with velocity $\mathbf{U} = U\mathbf{e}_z$ in a fluid at rest far from the sphere. We use a reference frame fixed with respect to the fluid at infinity and centred on the sphere at time t. We define spherical

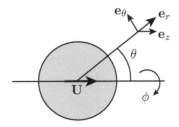

FIGURE 6.5
Sphere translating with velocity $\mathbf{U}$ in a fluid at rest at infinity.

coordinates with the z-axis in the direction of the particle velocity $\mathbf{U}$ (Figure 6.5). The flow is axisymmetric about Oz:

$$\partial/\partial\phi = 0$$

and the sphere does not rotate:

$$u_\phi = 0$$

The velocity field in the fluid is thus

$$\mathbf{u} = u_r(r,\theta)\mathbf{e}_r + u_\theta(r,\theta)\mathbf{e}_\theta$$

We have to solve the Stokes equations (6.3) with boundary conditions

$$\mathbf{u} \to 0 \quad \text{for} \quad r \to \infty \tag{6.39}$$

$$u_r = U\cos\theta \quad \text{and} \quad u_\theta = -U\sin\theta \quad \text{at} \quad r = a \tag{6.40}$$

6.3.1   Stream Function

The continuity equation in spherical coordinates reads as (Appendix B.2)

$$\frac{\partial}{\partial r}\left(r^2\sin\theta\, u_r\right) + \frac{\partial}{\partial\theta}\left(r\sin\theta\, u_\theta\right) = 0 \tag{6.41}$$

Here, the flow is not two-dimensional but three-dimensionnal and axisymmetric. It is however possible to introduce a stream function Ψ that allows to satisfy identically Equation (6.41):

$$u_r = \frac{1}{r^2\sin\theta}\frac{\partial\Psi}{\partial\theta} \quad \text{and} \quad u_\theta = \frac{-1}{r\sin\theta}\frac{\partial\Psi}{\partial r} \tag{6.42}$$

The condition at infinity (6.39) and the velocity expression (6.42) lead to

$$\Psi = o(r^2) \quad \text{for} \quad r \to \infty \tag{6.43}$$

where the notation $o(r^2)$ means of smaller order than r^2. The no-slip condition on the sphere (6.40) leads to

$$\Psi = \frac{1}{2}Ua^2\sin^2\theta \quad \text{and} \quad \frac{\partial\Psi}{\partial r} = Ua\sin^2\theta \quad \text{for} \quad r = a \tag{6.44}$$

In terms of Ψ, the Stokes equations for axisymmetric flows become

$$\mathfrak{D}^2\mathfrak{D}^2\Psi = 0 \tag{6.45}$$

where the operator $\mathfrak{D}^2$ is defined by

$$\mathfrak{D}^2 = \frac{\partial^2}{\partial r^2} + \frac{\sin\theta}{r^2}\frac{\partial}{\partial\theta}\left(\frac{1}{\sin\theta}\frac{\partial}{\partial\theta}\right) \tag{6.46}$$

Remark: $\mathfrak{D}^2$ is not the axisymmetric Laplacian operator.

We use the same method as the one presented in Chapter 3 for two-dimensional problems. Guided by the form of the boundary conditions (6.44), we seek a solution of the form

$$\Psi(r,\theta) = f(r)\sin^2\theta \tag{6.47}$$

A first application of the operator $\mathfrak{D}^2$ leads to

$$\mathfrak{D}^2\Psi = (f'' - 2f/r^2)\sin^2\theta = F(r)\sin^2\theta \tag{6.48}$$

where

$$F(r) = f'' - \frac{2f}{r^2} \tag{6.49}$$

We apply again the operator $\mathfrak{D}^2$:

$$\mathfrak{D}^2(\mathfrak{D}^2\Psi) = (F'' - 2F/r^2)\sin^2\theta = 0 \tag{6.50}$$

and seek a solution of the form $F = r^n$

$$F(r) = A'r^2 + C'/r \tag{6.51}$$

It follows from Equation (6.49) that

$$f(r) = Ar^4 + Br^2 + Cr + D/r \tag{6.52}$$

with $A = A'/10$ and $C = -C'/2$. The condition (6.43) leads to

$$A = B = 0$$

and conditions (6.44) lead to

$$f(a) = Ua^2/2, \qquad f'(a) = Ua$$

from which we find

$$C = \frac{3}{4}Ua \quad \text{and} \quad D = -\frac{1}{4}Ua^3 \tag{6.53}$$

The stream function is thus

$$\Psi = (Cr + D/r)\sin^2\theta = \frac{1}{4}Ua^2\left(\frac{3r}{a} - \frac{a}{r}\right)\sin^2\theta \tag{6.54}$$

We can write the stream function as the sum of two terms: the stokeslet and the potential dipole

$$\Psi = \Psi_S + \Psi_D$$

The stokeslet is defined by $\Psi_S = Cr\sin^2\theta$ and determines the flow field far from the particle (at 'infinity'). The potential dipole is defined by $\Psi_D = D\sin^2\theta/r$ and contributes to the flow field in the vicinity of the particle.

6.3.2   Velocity and Vorticity

The corresponding velocity field follows from Equations (6.42) and (6.54):

$$u_r = 2\left(\frac{C}{r} + \frac{D}{r^3}\right)\cos\theta \quad \text{and} \quad u_\theta = \left(-\frac{C}{r} + \frac{D}{r^3}\right)\sin\theta \tag{6.55}$$

The stokeslet dominates at infinity where the velocity is $O(1/r)$. This indicates that the perturbation created by a translating sphere extends far from the particle (for example, at 10 radii from the particle centre, the velocity is still 10% of U).

The flow vorticity vector is directed along $\mathbf{e}_\phi$

$$\boldsymbol{\omega} = \omega(r, \theta)\mathbf{e}_\phi$$

As for 2D flows, it is possible to write ω in terms of the stream function:

$$\omega = -\frac{\mathfrak{D}^2\Psi}{r\sin\theta} = \frac{2C}{r^2}\sin\theta \tag{6.56}$$

There is no contribution to vorticity from the potential dipole.

6.3.3   Force Exerted on the Sphere

A simple way to compute the pressure is to start from Equation (2.14):

$$\frac{\partial p}{\partial r} = -\mu(\nabla \times \boldsymbol{\omega})_r = -\frac{4\mu C}{r^3}\cos\theta \tag{6.57}$$

which is easily integrated

$$p - p_\infty = 2\mu C\cos\theta/r^2 \tag{6.58}$$

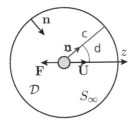

FIGURE 6.6
Flow domain $\mathcal{D}$ used to compute the force exerted on the sphere.

There are different ways to determine the resultant force $\mathbf{F}$ on the sphere. Starting with the expressions for p and $\mathbf{u}$, we use Newton's law to compute the stress tensor in the fluid. The integration of $\boldsymbol{\sigma} \cdot \mathbf{n}$ on the sphere yields the drag force

$$\mathbf{F} = \int_S \boldsymbol{\sigma} \cdot \mathbf{n} \, dS = \int_S \boldsymbol{\sigma} \cdot \mathbf{e}_r \, dS$$

It is possible though to simplify the computation. We first note that for a flow obeying Stokes equation

$$\int_{\mathcal{D}} \nabla \cdot \boldsymbol{\sigma} \, dV = 0$$

where $\mathcal{D}$ is the flow domain between the particle and another sphere S_∞ with radius $r_\infty \to \infty$, on which Equation (6.43) is satisfied (Figure 6.6). We then use Gauss' theorem

$$\int_{\mathcal{D}} \nabla \cdot \boldsymbol{\sigma} \, dV = \int_S \boldsymbol{\sigma} \cdot \mathbf{n} \, dS + \int_{S_\infty} \boldsymbol{\sigma} \cdot \mathbf{n} \, dS = 0$$

This means that instead of computing the resultant force on S, we can compute it on S_∞. The advantage is that the only contribution to the flow on S_∞ arises from the stokeslet, which simplifies the computation. Furthermore, we showed in Section 6.3 that the drag force was co-linear with the translation velocity. It follows that the force is simply $\mathbf{F} = F_z \mathbf{e}_z$ with

$$F_z = \int_{S_\infty} (\boldsymbol{\sigma} \cdot \mathbf{e}_r) \cdot \mathbf{e}_z \, dS = \int_{S_\infty} (\sigma_{rr} \cos\theta - \sigma_{r\theta} \sin\theta) \, dS \qquad (6.59)$$

The stokeslet stress tensor is

$$\sigma_{rr} = -p + 2\mu \, \partial u_r / \partial r = -p_\infty - 6C\mu \cos\theta / r^2$$
$$\sigma_{r\theta} = 2\mu e_{r\theta} = 0 \qquad\qquad\qquad\qquad\qquad\qquad\qquad (6.60)$$

Substituting Equation (6.60) into Equation (6.59), we get

$$F_z = -6C\mu \int_0^\pi \frac{\cos^2\theta}{r^2} 2\pi r^2 \sin\theta \, d\theta = -8\pi C\mu \qquad (6.61)$$

and replacing C with its value, we find

$$\mathbf{F} = -6\pi\mu a \mathbf{U} \tag{6.62}$$

which is the famous Stokes formula that relates the drag on a sphere to its translation velocity. We thus find that indeed $\lambda = 6\pi\mu$, as was announced in Section 6.2.

It is also possible to compute the local viscous traction $\boldsymbol{\sigma} \cdot \mathbf{n}$ on the sphere. From linearity considerations, we know that the stress is proportional to the cause which creates it, that is, the sphere velocity $\mathbf{U}$. Thus we can write

$$\boldsymbol{\sigma} \cdot \mathbf{n} = -\frac{\mu}{a} \mathbf{R}_T \cdot \mathbf{U} \tag{6.63}$$

where $\mathbf{R}_T$ is a dimensionless tensor which depends only on the particle geometry and which is isotropic when the particle is spherical (see Problem 6.6.4). The resultant force on the sphere is given by Equation (6.8)

$$\int_S \boldsymbol{\sigma} \cdot \mathbf{n} dS = -6\pi\mu a \mathbf{U} \tag{6.64}$$

From Equations (6.63) and (6.64) we find

$$\boldsymbol{\sigma} \cdot \mathbf{n} = -\frac{3\mu}{2a}\mathbf{U} \quad \text{and} \quad \mathbf{R}_T = \frac{3}{2}\mathbf{I} \tag{6.65}$$

This result can be verified by calculating the traction on the sphere from the stress created by the stokeslet and the potential dipole (the calculation is cumbersome, but straightforward).

6.3.4 Streamline Pattern

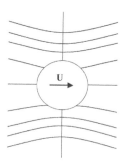

FIGURE 6.7
Streamline pattern around a sphere translating in a fluid at rest at infinity.

Streamline patterns around a sphere translating in a fluid at rest at infinity are shown in Figure 6.7. The upstream / downstream symmetry of the streamlines is due to the reversibility of Stokes flow (see Chapter 2, Problem 2.8.1).

6.4 Flow around a Rotating Sphere

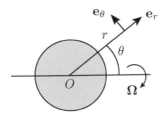

FIGURE 6.8
Sphere rotating with angular velocity Ω in a fluid at rest at infinity.

A sphere with radius a is rotating with angular velocity $\Omega = \Omega \mathbf{e}_z$ around its centre O, in a fluid at rest far from the sphere. We use a reference system linked to the fluid at infinity and centred on O at time t. Spherical coordinates are defined with z-axis along the direction of Ω. The velocity field is then of the form

$$\mathbf{u} = u_\phi(r, \phi)\mathbf{e}_\phi$$

We have to solve the Stokes equations (6.3) with boundary conditions (6.43) and with

$$u_\phi = \Omega a \sin\theta \quad \text{for} \quad r = a \tag{6.66}$$

We seek a solution of the form

$$u_\phi = \Omega a \, f(r) \sin\theta, \quad p' = p'_\infty \tag{6.67}$$

The continuity equation is identically satisfied by Equation (6.67) and the Stokes momentum equation becomes

$$\nabla^2 \mathbf{u} = \Omega a \sin\theta \left(f'' + \frac{2f'}{r} - \frac{2f}{r^2} \right) \mathbf{e}_\phi = 0 \tag{6.68}$$

with solution

$$f = Ar + B/r^2 \tag{6.69}$$

Condition (6.43) leads to $A = 0$, while condition (6.66) leads to $B = a^2$. The velocity and pressure fields around the rotating sphere are thus

$$\mathbf{u} = \Omega \frac{a^3}{r^2} \sin\theta \, \mathbf{e}_\phi, \quad p' = p'_\infty \tag{6.70}$$

In view of Equation (6.38), there is only a torque $G_z \mathbf{e}_z$ acting on the sphere

$$G_z = \int_0^\pi \sigma_{r\phi} \, a \sin\theta \, 2\pi a^2 \sin\theta \, d\theta \tag{6.71}$$

with

$$\sigma_{r\phi} = 2\mu e_{r\phi} = \mu \left[r \frac{\partial}{\partial r} \left(\frac{u_\phi}{r} \right) \right] = -3\mu\Omega \sin\theta$$

We find finally

$$\mathbf{G} = -8\pi\mu a^3 \, \Omega \tag{6.72}$$

which is the equivalent of Stokes formula for a rotating sphere. The resistance coefficient is $\eta = 8\pi$.

Note: We note that the determination of the two resistance coefficients λ and η is not straightforward and involves relatively heavy computations. We can then anticipate that the computations will be even more complicated for other geometries, even fairly simple ones such as ellipsoids [15, 16].

6.5   Slender Particles

6.5.1   Rod

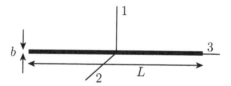

FIGURE 6.9
Straight slender rod or fibre with length L directed along Ox_3 and transverse dimension $b \ll L$.

The length L of a straight slender particle (fibre, rod) is much larger than the other transversal dimensions b, $b \ll L$. In this case, the characteristic length of the flow will be L (the proper scale for the hydrodynamic perturbations created by a rotating rod) rather than a (the radius of the sphere with the same volume as the rod). For a cylindrical rod, the eigenvalues of the resistance tensors are equal along directions 1 and 2 (Figure 6.9). When the rod is slender, $L \gg b$, the eigenvalues of the resistance tensor are given to first order by [4]:

$$\lambda_3 \cong \frac{2\pi}{\ln(2L/b)} \cong \frac{\lambda_1}{2} \cong \frac{\lambda_2}{2} \tag{6.73}$$

The drag force on a slender rod or fibre will thus be

$$\mathbf{F} = \frac{-4\pi\mu L}{\ln(2L/b)} \{ \ U_1, \quad U_2, \quad \tfrac{1}{2}U_3 \ \} \tag{6.74}$$

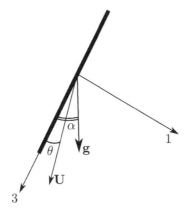

FIGURE 6.10
Slender rod falling under gravity in a fluid at rest. When the rod makes an angle α with gravity, its velocity is not vertical.

When a slender rod is falling under gravity, its velocity is not vertical. Indeed, for a rod making an angle α with respect to gravity (Figure 6.10), the external force is

$$\mathbf{F}_e = m'g\cos\alpha\,\mathbf{e}_3 + m'g\sin\alpha\,\mathbf{e}_1$$

where m' is the mass of the rod corrected with Archimedes force. The hydrodynamic drag is

$$\mathbf{F} = \frac{-4\pi\mu L}{\ln(2L/b)}\left(\frac{1}{2}U\cos\theta\,\mathbf{e}_3 + U\sin\theta\,\mathbf{e}_1\right)$$

where θ is the angle between $\mathbf{U}$ and the rod axis $\mathbf{e}_3$. Since the two forces are equal, we find

$$\tan\alpha = 2\tan\theta$$

The rod thus migrates in the lateral direction while it falls.

6.5.2 Helix

A circular helix with axis Oz, radius a and pitch $2\pi p$ is translating with velocity $\mathbf{U} = U\mathbf{e}_z$ in a fluid at rest at infinity (Figure 6.11). The helix thickness is much smaller than a. We consider the flow near a small helix element $ds\,\boldsymbol{\tau}$ where $\boldsymbol{\tau}$ is the unit tangent vector

$$ds\,\boldsymbol{\tau} = -a\sin\theta\,d\theta\mathbf{e}_x + a\cos\theta d\theta\,\mathbf{e}_y + pd\theta\,\mathbf{e}_z \tag{6.75}$$

centred on point M with coordinates

$$\mathbf{OM} = a\cos\theta\,\mathbf{e}_x + a\sin\theta\,\mathbf{e}_y + p\theta\,\mathbf{e}_z \tag{6.76}$$

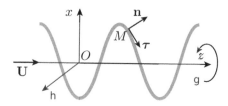

FIGURE 6.11
Helix translating with velocity $\mathbf{U} = U\mathbf{e}_z$ in a fluid at rest.

The small element can be treated as a slender body of length ds. It is subjected to a drag force $\mathbf{dF}$

$$d\mathbf{F} = -\lambda\mu ds\, U_t \boldsymbol{\tau} - 2\lambda\mu ds\, U_n \mathbf{n} \qquad (6.77)$$

where λ is the resistance coefficient of a rod for longitudinal motion, and U_t and U_n are the velocity components tangent and normal to the helix, respectively,

$$\mathbf{U} = U\mathbf{e}_z = U_t \boldsymbol{\tau} + U_n \mathbf{n}$$

Eliminating U_n from Equation (6.77), we find the drag on the helix element:

$$d\mathbf{F} = \lambda\mu ds\, U_t \boldsymbol{\tau} - 2\lambda\mu ds\, U\mathbf{e}_z \qquad (6.78)$$

The axial component G_z of the torque due to the drag force is

$$dG_z = [\mathbf{OM} \times (\lambda\mu ds\, U_t \boldsymbol{\tau} - 2\lambda\mu ds\, U\mathbf{e}_z)] \cdot \mathbf{e}_z = \frac{\lambda\mu pa^2 U\, d\theta}{\sqrt{a^2 + p^2}} \qquad (6.79)$$

The total torque per coil is then

$$G_z = \frac{2\pi\lambda\mu pa^2 U}{\sqrt{a^2 + p^2}} \qquad (6.80)$$

This torque being positive means that the helix rotates in the trigonometric direction as it moves along Oz. Conversely, rotation of the helix around its axis Oz creates an axial propulsion force. Numerous micro-organisms (e.g. *Escherichia coli*) use this principle to swim: they rotate a flagellum by means of muscular engines and thus obtain a propulsion force.

6.6   Problems

6.6.1   Sedimentation of a Particle with a Symmetry Plane

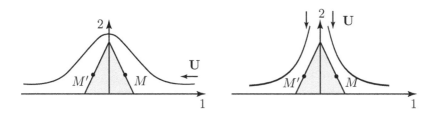

FIGURE 6.12
Solid particle with symmetry plane $x_1 = 0$, subjected to two different flows. The points M and M' are symmetrical with respect to the plane of symmetry.

A solid particle with a plane of symmetry moves in a fluid at rest at infinity. Use a reference frame linked to the fluid at infinity and centred on the particle in such a way that $x_1 = 0$ is the plane of symmetry (Figure 6.12).

1. Determine which components of the resistance tensors are non-zero. Hint: Consider two different particle translations with no rotation $\mathbf{U} = U_1 \mathbf{e}_1$ and $\mathbf{U} = U_2 \mathbf{e}_2$.

2. Apply the result to a spherical particle and find the general form of the resistance tensors.

3. Show that if the particle is oriented so that gravity $\mathbf{g}$ is normal to the plane of symmetry, its motion will be such that

$$U_2 = U_3 = \Omega_1 = 0$$

6.6.2   Swimming Motion of Spheres and Bacteria

From an Ecole Polytechnique problem written with E. Guazzelli

The aim of this problem is to study the motion of a spherical particle due to a velocity field on its surface. The second part of the problem deals with the creation of such a velocity field with an application to the swimming motion of micro-organisms.

A spherical particle with radius a is suspended in a Newtonian incompressible liquid (viscosity μ and density ρ) at rest far from the particle. We neglect gravity and thus no external force acts on the particle. We use a spherical coordinate system (r, θ, ϕ) centred on the sphere and assume that the flow Reynolds number around the sphere is very small.

1. In a reference frame linked to the sphere, a velocity field $\mathbf{u}' = K \sin\theta\, \mathbf{e}_\theta$ is applied on the surface S of the sphere. This leads to a bulk motion of the sphere with velocity $U(t)\mathbf{e}_z$. Consequently, in a reference frame linked to the fluid at rest at infinity, the velocity $\mathbf{u}_s$ of a point of the surface S of the sphere is given by

$$\mathbf{u}_s = U(t)\mathbf{e}_z + \mathbf{u}' = U(t)\mathbf{e}_z + K \sin\theta\, \mathbf{e}_\theta \qquad (6.81)$$

   (a) Give the boundary conditions for the stream function $\Psi(r,\theta)$ of the flow around the sphere and find the general form of Ψ.
   (b) Compute Ψ.
   (c) Find the propulsion velocity $\mathbf{U}$ of the sphere.

2. The objective is now to find the propulsion velocity due to a general surface velocity field $\mathbf{u}_s$.

   We first consider a Stokes flow (T) where the sphere is subjected to an external force $\mathbf{F}^T$ which results into a translation motion with velocity $\mathbf{U}^T$ in a fluid at rest. The local velocity and stress fields in the fluid are denoted $\mathbf{u}^T$ and $\boldsymbol{\sigma}^T$.

   We then consider a second Stokes flow (SW) where the sphere is force free and moves with velocity $\mathbf{U}(t)$ in a fluid at rest, under the effect of a surface velocity field $\mathbf{u}_s = \mathbf{U}(t) + \mathbf{u}'(\mathbf{x},t)$ for $\mathbf{x} \in S$. The corresponding local velocity and stress fields in the fluid are denoted $\mathbf{u}$ and $\boldsymbol{\sigma}$.

   (a) Give the boundary conditions for flow (SW).
   (b) In a clearly defined fluid domain, apply the reciprocal theorem to flows (T) and (SW) and infer the relation between $\mathbf{U}(t)$ and $\mathbf{u}'$.
   (c) We note $\langle\ \rangle$ the average over a time interval T of the quantity in brackets. Show that

$$\langle \mathbf{U}(t) \rangle = \frac{-1}{4\pi a^2} \int_S \langle \mathbf{u}'(\mathbf{x},t) \rangle dS$$

3. We now study the generation of the surface velocity field $\mathbf{u}'$. Nature does it by providing micro-organisms with a deformable membrane or with a system of surface cilia. In both cases, a deformation wave propagates on the surface of the bacteria, either directly on the membrane or on the tip of the cilia. The motion of the surface S is then described by means of a Lagrangian approach, whereby a material point of the surface is identified by its spherical coordinates a, θ_0, ϕ_0. At time t, this point is moved to position a, $\theta = \theta(\theta_0,t)$, $\phi = \phi_0$ with

$$\theta = \theta_0 + \varepsilon \cos(n\theta_0 - \omega t)$$

where ε measures the wave amplitude ($\varepsilon \ll 1$). The time average $\langle\ \rangle$ is now computed over a period $T = 2\pi/\omega$.

(a) Determine $\mathbf{u}'$ and $\langle \mathbf{U}(t) \rangle$ to first order in ε.

(b) Determine $\mathbf{u}'$ and $\langle \mathbf{u}' \rangle$ to second order in ε.

(c) Find the average bulk velocity of the particle $\langle \mathbf{U} \rangle$.

(d) For a micro-organism velocity of order $10\, a\ \mathrm{s}^{-1}$, $\varepsilon = 1/20$ and $n = 5 - 10$, what would be the maximum size of the micro-organism for the analysis to be valid?

4. As a bonus, design a micro-robot that can swim as in this problem.

6.6.3 Micro-Organism Propulsion by Means of a Flagellum

From an Ecole Polytechnique problem written with E. Guazzelli

FIGURE 6.13
Micro-organism configuration at time t. The flagellum has an oscillatory motion in plane xy. The full line represents the flagellum and the dashed line shows the wave shape.

Large living creatures swim in water by means of an oscillatory motion of their body, using the inertia of the generated fluid flow. Some microscopic creatures such as spermatozoids also swim by means of an oscillatory motion of a flagellum. The two processes may seem identical, but rely on quite different physical effects. Indeed, the swimming motion of a micro-organism generates a flow with a very small Reynolds number (based on the organism's dimension). Consequently, inertia effects are negligible and the propulsion mechanism is fundamentally different from that of large organisms. The aim of this problem is to study how a micro-organism can swim by oscillating a flagellum and beat the reversibility of Stokes flows. This problem has application in the propagation of bacteria with flagella or mammal reproduction (sterility diagnostic or artificial insemination *in vitro*).

The micro-organism consists of a head and of an inextensible flagellum of length L and mean thickness b, such that $b \ll L$. Internal muscular engines can move the flagellum.

The micro-organism swims in a fluid at rest (viscosity μ). Gravity is negligible, and since $b \ll L$, the flagellum can be treated as a slender body. We use a reference frame linked to the fluid at infinity and centred at time t on the junction between the head and the flagellum (Figure 6.13). The micro-organism moves in the fluid with a constant velocity $\mathbf{U} = -U\mathbf{e_x}$, where $U \geqslant 0$. We make the simplifying assumption that the flagellum motion is planar. Thus the position of a point M of the flagellum, with coordinates (x, y) at time t, is given by an oscillatory motion in the xy-plane:

$$y = Y(x - Vt)$$

where the function Y defines the shape of the wave and where V is the wave velocity $(V > U)$. The unit vectors tangent and normal to the flagellum at point M at time t are denoted $\boldsymbol{\tau}$ and $\mathbf{n}$, respectively. The vectors $\boldsymbol{\tau}$ and $\mathbf{e_x}$ make and angle θ (Figure 6.13)

$$\boldsymbol{\tau} = \cos\theta\mathbf{e_x} + \sin\theta\mathbf{e_y}$$

1. Motion at time t

   (a) The velocity of a point M of the flagellum is

   $$\mathbf{u} = u\mathbf{e_x} + v\mathbf{e_y} = u_t\boldsymbol{\tau} + u_n\mathbf{n}$$

   Compute the components u, v, u_t and u_n in terms of U, V and θ.

   (b) The flagellum is treated as a slender body with constant resistance coefficients λ_1 and λ_3 (Section 6.5). Compute the force $\mathbf{dF}$ exerted by the fluid on a small element of flagellum $d\ell\boldsymbol{\tau}$ centred on point M.

   (c) We define

   $$\int_0^L \cos^2\theta\, d\ell = \beta L, \qquad \int_0^L \cos^{-2}\theta\, d\ell = \beta' L$$

   Show that the x-component of the total force exerted by the fluid on the flagellum is

   $$F_x = -\lambda_1\mu L[\gamma\beta(V - U) - \gamma V + (V - U)(1 - \beta)]$$

   where $\gamma = \lambda_3/\lambda_1$.

   (d) The wave motion of the flagellum moves the head of the micro-organism with a velocity $\mathbf{U}$. We assume that the head is roughly spherical with radius a ($a \ll L$ and $a \gg b$). Compute the velocity ratio U/V in terms of β, γ and $\delta = 6\pi a/\lambda_1 L$. Discuss the validity of the hypotheses and of the approximations.

   (e) The organism is stationary (for example, the head is fixed on a wall). Compute the thrust F_0 generated by the oscillatory motion of the flagellum.

(f) Compute the rate of mechanical work dP that the micro-organism must produce to move a flagellum element $d\ell\boldsymbol{\tau}$ with velocity $\mathbf{u}$. Compute the total rate of mechanical work P for the whole flagellum and discuss the result.

(g) Some micro-organisms (e.g. spirochetes) have only a flagellum with no head. Compute the swimming velocity U of such micro-organisms as a function of the wave velocity V on the flagellum.

2. Effect of time and discussion

Typical values for cattle spermatozoid propulsion are $a = 5\,\mu$m, $L = 60\,\mu$m, $b = 0.2\,\mu$m, $U/V = 0.1$, $h = 8\,\mu$m, $L/\Lambda = 0.5$, $\mu = 10^{-3}$ Pa s where h and Λ are defined in Question 2.b.

(a) Discuss the validity of the slender body hypothesis. Estimate the value of γ.

(b) As an example, we consider a sinusoidal wave with amplitude h and wave-length Λ such that $\Lambda \gg L$:

$$y = h\sin\left[\frac{2\pi}{\Lambda}(x - Vt)\right]$$

Compute $\beta(t)$ to first order in L/Λ.

(c) In the case of a freely swimming spermatozoid (Question 1.d), describe the time evolution of U. What is the value of U for $\beta = 1$? What are the maximum and minimum values of U? How is it possible to increase the maximum value of U?

(d) Draw on the same graph a sketch of the positions and shapes of the spermatozoid at times t, $t + \frac{\Lambda}{4V}$, $t + 2\frac{\Lambda}{4V}$, $t + 3\frac{\Lambda}{4V}$.

6.6.4 Solid Particle in a General Flow Field

From an Ecole Polytechnique problem written with E. Guazzelli

A solid particle is fixed in a fluid undergoing Stokes flow with velocity $\mathbf{u}^\infty(\mathbf{x})$ and pressure $p^\infty(\mathbf{x})$ far away from the particle (Figure 6.14). The particle has a characteristic dimension a, its surface is noted S with outer unit normal vector $\mathbf{n}$. The aim of the problem is to compute the resultant force $\mathbf{F}$ that must be exerted on the particle to keep it motionless or, conversely, to compute the velocity of a particle freely suspended in a general shear flow.

The presence of the particle perturbs the flow. Consequently, the velocity and pressure near the particle can be written as

$$\begin{aligned} \mathbf{u}(\mathbf{x}) &= \mathbf{u}^\infty(\mathbf{x}) + \mathbf{u}^D(\mathbf{x}) \\ p(\mathbf{x}) &= p^\infty(\mathbf{x}) + p^D(\mathbf{x}) \end{aligned}$$

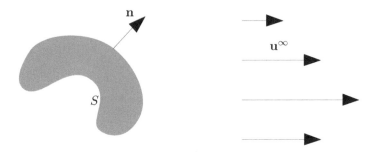

FIGURE 6.14

Motionless solid particle in an arbitrary flow.

1. Give the equations of the perturbed flow $\mathbf{u}^D(\mathbf{x})$, $p^D(\mathbf{x})$ together with the corresponding boundary conditions.

2. It is possible to write the resultant force on the particle as $\mathbf{F} = \mathbf{F}^\infty + \mathbf{F}^D$. Compute $\mathbf{F}^\infty$ and $\mathbf{F}^D$ in terms of the corresponding stress tensors. Show that $\mathbf{F}^\infty = 0$.

3. We consider now a second Stokes flow in which the particle is moving with constant velocity $\mathbf{U}$ in the same fluid that is now at rest at infinity. We note $\mathbf{u}^T(\mathbf{x})$, $p^T(\mathbf{x})$ the velocity and pressure fields created by this motion of the particle, respectively. Show that the surface density $\mathbf{f}^T$ of the force exerted by the fluid on the particle is

$$\mathbf{f}^T = \boldsymbol{\sigma}^T \cdot \mathbf{n} = -\frac{\mu}{a} \mathbf{R_T} \cdot \mathbf{U}$$

where $\mathbf{R_T}$ is a dimensionless tensor. Find the relation between $\mathbf{R_T}$ and the resistance tensor $\mathbf{A}$.

4. Apply the reciprocal tensor to the two flows $\mathbf{u}^D(\mathbf{x})$ and $\mathbf{u}^T(\mathbf{x})$ in a properly defined domain and compute the product $\mathbf{U} \cdot \mathbf{F}^D$ in terms of $\mathbf{u}^D$ and $\mathbf{f}^T$ evaluated on S.

5. Show that the force that must be exerted on the particle to keep it motionless is given by

$$\mathbf{F} = \frac{\mu}{a} \int_S \mathbf{u}^\infty \cdot \mathbf{R_T} \, dS$$

6. The particle is now a sphere with radius a. Compute $\mathbf{R_T}$ and the force that must be applied to keep the sphere motionless in the flow $\mathbf{u}^\infty(\mathbf{x})$.

7. The sphere is now freely suspended (no external force or torque) in the general flow field $\mathbf{u}^\infty(\mathbf{x})$ and has a constant velocity $\mathbf{V}$. Compute $\mathbf{V}$ in terms of $\mathbf{u}^\infty(\mathbf{x})$.

7

Flow of Bubbles and Droplets

CONTENTS

In this chapter we consider the flow of *deformable* particles suspended inside a liquid. The typical case is that of an emulsion of two immiscible liquids (oil in water or air in water, for example). The objective is to compute the motion and the eventual deformation of the droplets or bubbles as they move in the fluid and are subjected to hydrodynamic forces. This eventually leads to the determination of the velocity of phase separation and to the effective transport properties of such two-phase systems. Obviously the problem is complicated by the presence of a deformable interface of unknown geometry on which important boundary conditions are expressed. Different techniques are available to solve the Stokes equations in this case:

- For axisymmetric situations, spherical coordinates and a stream function can be used.

- For problems with a simple geometry (spherical or quasi-spherical), it is possible to use Cartesian coordinates and a singularity method (Chapter 8).

- For problems with a complex geometry, a numerical model based on the integral formulation of the Stokes equation is usually very efficient (Chapter 8).

7.1 Freely Suspended Liquid Drop

7.1.1 General Problem Statement

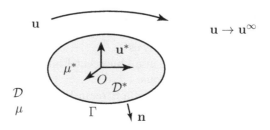

FIGURE 7.1
Liquid droplet (domain $\mathcal{D}^*$) suspended in another immiscible liquid (domain $\mathcal{D}$) subjected to the flow $\mathbf{u}^\infty$ far from the drop.

We first present the general formulation of the flow around a suspended droplet. Some specific examples are then given and solved.

A liquid droplet with viscosity μ^* and density ρ^* is suspended in another liquid with viscosity μ and density ρ. We denote $\mathcal{D}^*$ and $\mathcal{D}$ the fluid domains corresponding to the drop and to the surrounding liquid, respectively. At rest, the drop is spherical with radius a. We use a reference system linked to the ambient fluid and centred on the drop centre of mass at time t. Far from the drop, the fluid velocity is $\mathbf{u}^\infty$ and the pressure is p^∞. We assume that the Reynolds number of the flow about the drop is very small:

$$Re = \rho \left| \mathbf{u}^\infty \right| a / \mu \ll 1$$

The smaller the drop, the easier it will be to fulfill this condition. The flows around and inside the drop thus satisfy the Stokes equations

$$\nabla \cdot \mathbf{u} = 0 \quad \text{and} \quad \mu \nabla^2 \mathbf{u} - \nabla p = 0 \quad \text{for} \quad \mathbf{x} \in \mathcal{D} \tag{7.1}$$

$$\nabla \cdot \mathbf{u}^* = 0 \quad \text{and} \quad \mu^* \nabla^2 \mathbf{u}^* - \nabla p^* = 0 \quad \text{for} \quad \mathbf{x} \in \mathcal{D}^* \tag{7.2}$$

where the gravity forces have been added to the pressure terms (Chapter 1). A first boundary condition is that the perturbation due to the presence of the drop dies out far from the drop:

$$\mathbf{u} \to \mathbf{u}^\infty \quad \text{for} \quad |\mathbf{x}| \to \infty \tag{7.3}$$

The other boundary conditions are expressed on the free surface Γ between the two liquids. The unknown equation of Γ can be written as

$$F(x_1, x_2, x_3, t) = 0 \tag{7.4}$$

where the function F has to be determined. The free-surface boundary conditions are then (Chapter 5)

- The continuity of velocity

$$\mathbf{u}(\mathbf{x}) = \mathbf{u}^*(\mathbf{x}) \tag{7.5}$$

- The kinematic condition which expresses the impermeability of Γ

$$\frac{1}{|\nabla F|}\frac{\partial F}{\partial t} + \mathbf{u} \cdot \mathbf{n} = 0 \tag{7.6}$$

where $\mathbf{n}$ is the unit normal vector to Γ pointing out of the drop (Figure 7.1)

- The dynamic condition for the *hydrodynamic* stress jump

$$(\boldsymbol{\sigma}' - \boldsymbol{\sigma}'^*) \cdot \mathbf{n} = \gamma_s \kappa \mathbf{n} \tag{7.7}$$

where κ is the interface curvature and γ_s the surface tension between the two liquids, which is assumed to remain constant along the interface.

The solution to this set of equations allows us to determine the shape of the droplet and the relation between the drop velocity, the external flow and the external forces that act on the drop. However, this problem is very difficult to solve analytically unless the drop shape remains nearly spherical.

7.1.2 Dimensional Analysis

A dimensional analysis of the problem reveals that the main dimensionless numbers are (apart from the Reynolds number which is assumed to be negligible):

- The viscosity ratio $\lambda = \mu^*/\mu$ between the drop and the ambient liquid
- The capillary number $Ca = \mu |\mathbf{u}^\infty| /\gamma_s$ which measures the ratio between the viscous forces and the surface tension.

When λ is large, the viscous droplet will not deform much. In the limit $\lambda \gg 1$, the liquid drop behaves like a solid sphere. Similarly, when $Ca \ll 1$, the surface tension forces (which tend to minimise the interfacial area and thus to make the drop spherical) are much larger than the viscous deforming forces. Thus in the asymptotic case $Ca \ll 1$, the droplet deformation is small and the drop is quasi-spherical. It is then possible to seek a solution to Equations (7.1) to (7.7) by means of the singularity expansion technique (Chapter 8) and successive approximations [1, 24]. However, the analytical solution becomes quickly very difficult to manipulate by hand.

In the case of the sedimentation of a liquid droplet, the velocity scale is given by $\Delta \rho\, a^2\, g/\mu$ (Chapter 6, Section 6.2), where $\Delta \rho = |\rho^* - \rho|$. Then the capillary number becomes the Bond number

$$Ca = \Delta \rho\, a^2 g/\gamma_s = B \tag{7.8}$$

which measures the ratio between the gravity and capillary forces (Chapter 5, Section 5.2.2). If $B \ll 1$, the drop remains quasi-spherical when it sediments.

7.2   Translation of a Bubble in a Quiescent Fluid

We now solve the previous problem in the simple case of a gas bubble sedimenting in a fluid at rest with constant velocity. This problem coupled with mass transport equations, is important for the oxygenation of tanks (reaction tanks or fish tanks) or for the organoleptic quality of sparkling beverages.

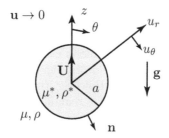

FIGURE 7.2
Gas bubble moving with velocity $\mathbf{U}$ in a fluid at rest.

7.2.1   Problem Equations

A spherical bubble with radius a moves in a fluid at rest under the effect of gravity. At steady state, the bubble velocity is $\mathbf{U}$ in a reference frame linked to the liquid and centred on the bubble at time t. The liquid viscosity and density are μ and ρ, respectively. The bubble viscosity and density are both small with respect to those of the liquid:

$$\rho^* \ll \rho \qquad \text{and} \qquad \mu^* \ll \mu$$

We assume that Ca and B are very small ($Ca \ll 1$ and $B \ll 1$) and thus that the bubble is almost spherical (we shall verify later the validity of these hypotheses). Equations (7.1) to (7.7) become

$$\nabla.\mathbf{u} = 0 \quad \text{and} \quad \mu\nabla^2\mathbf{u} - \nabla p = 0 \quad \text{for} \quad \mathbf{x} \in \mathcal{D} \tag{7.9}$$

$$\nabla p^* = 0 \quad \text{for} \quad \mathbf{x} \in \mathcal{D}^* \tag{7.10}$$

with boundary conditions

- From Equation (7.3),

$$\mathbf{u} \to 0 \quad \text{for} \quad |\mathbf{x}| \to \infty \tag{7.11}$$

- From Equation (7.6),

$$\mathbf{u}(\mathbf{x}) \cdot \mathbf{n} = \mathbf{U} \cdot \mathbf{n} \quad \text{for} \quad \mathbf{x} \in \Gamma \tag{7.12}$$

- From Equation (7.7),
  - Tangential stress component

$$\boldsymbol{\sigma} \cdot \mathbf{n} - (\mathbf{n} \cdot \boldsymbol{\sigma} \cdot \mathbf{n})\mathbf{n} = 0 \quad \text{or} \quad \sigma_{ik}n_k - (\sigma_{km}n_kn_m)n_i = 0 \tag{7.13}$$

  - Normal stress component

$$\mathbf{n} \cdot \boldsymbol{\sigma}'\mathbf{n} - (-p'^*) = \gamma_s\kappa \quad \text{or} \quad \sigma'_{km}n_kn_m - (-p'^*) = \gamma_s\kappa \tag{7.14}$$

where p' and p'^* represent the hydrodynamic pressures (with no hydrostatic effect included) in each fluid. The condition on the tangential stress (7.13) has no contribution from the isotropic terms and thus applies indifferently to the hydrodynamic or to the modified stress. The condition (7.5) is meaningless here because we do not try to compute the flow inside the bubble in view of the negligible viscosity.

7.2.2 Solution in Terms of a Stream Function

We follow the same solution technique as the one presented for a solid sphere undergoing steady translation in a quiescent liquid (Chapter 6, Section 6.3). We use a spherical coordinate system with z-axis along the translation velocity $\mathbf{U}$. The problem is then axisymmetric around this direction (Figure 7.2). Note that it is also possible to use the singularity method and keep a Cartesian coordinate system (see Chapter 8, Section 8.3.3).

In the spherical coordinate system, the boundary conditions (7.12) to (7.14) become on $r = a$

$$u_r = U\cos\theta \quad \text{and} \quad \sigma_{r\theta} = 2\mu e_{r\theta} = 0 \tag{7.15}$$

$$\sigma'_{rr} + p'^* = \gamma_s\kappa \tag{7.16}$$

The stream function Ψ has the same form as for a solid sphere (Equation (6.47))

$$\Psi(r,\theta) = f(r)\sin^2\theta$$

It satisfies the same bi-harmonic equation (6.45) with the same boundary condition at infinity (6.43). Without redoing all the computations, we thus know that the unique solution is the sum of a stokeslet and of a potential dipole

$$\Psi = (Cr + D/r)\sin^2\theta \tag{7.17}$$

with associated velocity field

$$u_r = 2\left(\frac{C}{r} + \frac{D}{r^3}\right)\cos\theta, \quad u_\theta = \left(-\frac{C}{r} + \frac{D}{r^3}\right)\sin\theta \tag{7.18}$$

The expressions of u_r and u_θ allow us to compute the rate of strain

$$e_{r\theta} = -3D\sin\theta/r^4 \tag{7.19}$$

The boundary conditions (7.15) then lead to

$$C = Ua/2 \quad \text{and} \quad D = 0 \tag{7.20}$$

The flow around the bubble is only a stokeslet.

Following again the same reasoning as for a solid sphere (Chapter 6, Section 6.3.3), we find that the drag force is parallel to $\mathbf{U}$ and given by

$$F_z = -8\pi C\mu \quad \text{or} \quad \mathbf{F} = -4\pi\mu a\mathbf{U} \tag{7.21}$$

which is the equivalent of the Stokes formula for an inviscid bubble. The rise velocity of the bubble follows from the force balance

$$\mathbf{F} + \frac{4}{3}\pi a^3(\rho^* - \rho)\mathbf{g} = 0 \tag{7.22}$$

or

$$\mathbf{U} = (\rho^* - \rho)a^2\mathbf{g}/3\mu \cong -\rho a^2\mathbf{g}/3\mu \tag{7.23}$$

The bubble moves upwards with a velocity which is proportional to the square of its radius, in a fashion similar to a solid sphere.

7.2.3 Shape of the Bubble

In fact, the bubble can deform and may not remain spherical. The shape of the bubble is determined by the competition between the deforming viscous traction due to the motion of the ambient liquid and the capillary force which tends to minimise the interfacial area and thus keep the bubble spherical. We must now verify that a spherical bubble shape does satisfy all the problem equations. The only equation that involves the bubble geometry (through the radius of curvature of the surface) is the normal stress condition (7.16).

In principle, to find the bubble shape, we first compute κ from Equation (7.16), and then integrate it to find the surface equation (a difficult problem in general!).

Accordingly, we first compute the normal stress jump across the interface and do not forget that Equation (7.16) is written for the hydrodynamic stress. Equation (7.10) shows that the modified pressure (including gravity) is constant in $\mathcal{D}^*$

$$p^* = p_0^* \quad \text{for} \quad \mathbf{x} \in \mathcal{D}^* \tag{7.24}$$

The hydrodynamic pressure inside the bubble is thus

$$p'^* = p_0^* - \rho^* ga \cos\theta \tag{7.25}$$

In the suspending liquid, the contribution of the stokeslet to the modified normal stress σ_{rr} is given in Chapter 6, Section 6.3.3

$$\sigma_{rr} = -p_0 - 6C\cos\theta/a^2 = -p_0 - 3\mu U \cos\theta/a \tag{7.26}$$

where p_0 is constant. The hydrodynamic normal stress in the liquid is thus

$$\sigma'_{rr} = -p_0 + \rho ga \cos\theta - 3\mu U \cos\theta/a \tag{7.27}$$

The condition (7.16) on the normal stress then becomes

$$-p_0 + \rho ga \cos\theta - 3\mu U \cos\theta/a + p_0^* - \rho^* ga \cos\theta = \gamma_s \kappa \tag{7.28}$$

Replacing U by its value (7.23), we find that the viscous term exactly balances the hydrostatic term. We are then left with

$$-p_0 + p_0^* = \gamma_s \kappa \tag{7.29}$$

which indicates that the curvature of the interface has the same value for all points, that is, the bubble is spherical! This result is obtained for any value of the Bond number. We conclude that a spherical bubble is a valid solution of the problem (but is it stable?... see Section 7.3.2 and Section 7.3.3).

7.3 Translation of a Liquid Drop in a Quiescent Fluid

7.3.1 Hadamard–Rybczynski Drag on a Drop

We now consider the flow around a liquid droplet moving with a constant velocity $\mathbf{U}$ in an immiscible liquid at rest at infinity. The situation is very similar to the one studied in the preceding section. The main difference is that the drop viscosity μ^* is no longer negligible with respect to the ambient viscosity μ. This means that the internal flow inside the drop must be determined as

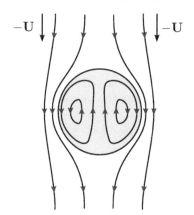

FIGURE 7.3
Streamlines around and inside a liquid droplet moving with velocity **U** in a
liquid at rest at infinity. The streamlines are shown in a reference frame linked
to the drop.

part of the solution. We assume again that Ca and B are small ($Ca \ll 1$
and $B \ll 1$) and that the droplet is spherical with radius a. We use spherical
coordinates with axis parallel to the drop velocity.

We must now solve Equations (7.1) and (7.2) (instead of Equation (7.10)).
The boundary conditions are then given by Equations (7.11), (7.12) and (7.7),
to which we must add Equation (7.5). The same reasoning as for a solid sphere
or a gas bubble indicates that the stream function Ψ of the external flow is
the sum of a stokeslet and of a potential dipole

$$\Psi = (Cr + D/r)\sin^2\theta$$

Inside the droplet, we seek a stream function Ψ^* of the form

$$\Psi^* = f^*(r)\,sin^2\theta$$

which allows us to satisfy the boundary conditions on the interface. The solution of the bi-harmonic equation then leads to a solution with the same form
as Equation (6.52)

$$f^*(r) = A^* r^4 + B^* r^2 + C^* r + D^*/r \tag{7.30}$$

Since the velocity must be finite when $r = 0$, we take $C^* = D^* = 0$. We then
use the boundary conditions to compute the coefficients A^*, B^*, C and D, and
find (after some computations)

$$C = Ua\frac{3\lambda + 2}{4(\lambda + 1)}, \qquad D = -Ua^3\frac{\lambda}{4(\lambda + 1)},$$

$$B^* = U\frac{2\lambda + 3}{4(\lambda + 1)}, \qquad A^* = -\frac{U}{a^2}\frac{1}{4(\lambda + 1)}$$

The corresponding stream lines are shown in Figure 7.3 where they are computed in a reference frame linked to the droplet. Note the apparition of two recirculation torus inside the drop.

As previously, the drag on the droplet is given by the stokeslet coefficient

$$\mathbf{F} = 8\pi\mu C\mathbf{e}_z = -2\pi\mu a \frac{3\lambda + 2}{\lambda + 1}\mathbf{U} \qquad (7.31)$$

This result is known as the Hadamard–Rybczynski formula. We note that we recover the Stokes formula (6.62) when $\lambda \to \infty$, and the gas bubble formula (7.21) when $\lambda = 0$. It is then easy to compute the terminal settling velocity of the liquid droplet.

When we compute the normal stress jump across the interface, we find again that it is constant and equal to the pressure difference across the interface. We conclude that the spherical shape hypothesis allows us to satisfy all the problem equations for any value of the surface tension. This result is surprising and counter intuitive since we would expect that a liquid drop with a very low surface tension could easily deform. Indeed the solutions we just found are not always stable.

7.3.2 Stability of the Stokes Solution

The Stokes equations correspond to a balance between the viscous forces and the pressure forces at each time. The time stability of a solution of a Stokes problem is thus not guaranteed. Indeed, the sedimentation of droplets has been modelled numerically for different values of the Bond number and of the viscosity ratio λ in the Stokes regime [33]. The initial shape of the droplet is slightly perturbed with respect to the sphere. If a spherical shape is a stable solution of the flow, then the drop shape must return to a sphere as the sedimentation proceeds. The numerical results show that this happens only for 'small' values of the Bond number (of order 10 or less). For large values of the Bond number, the geometrical perturbation increases with time. Depending on the initial perturbation, we can obtain drops with a tail or drops with a central dimple (Figure 7.4). These numerical findings have been confirmed by experiments (Figure 7.5).

7.3.3 Validity Limits of the Stokes Solution

The solutions we obtained for sedimenting drops or bubbles are valid inasmuch as the flow satisfies the condition $Re \ll 1$. For a gas bubble with radius 0.2 mm in water ($\Delta\rho = 10^3$ kg/m^3, $\mu = 10^{-3}$ Pa s, $\gamma_s = 0.07$ N/m), the sedimentation velocity is $U \cong 0.4$ m/s and $Re \cong 80$, which is not small. The computation of Section 7.2 applies then for bubbles with a radius of a few tens of microns. However, experiments show that the bubble remains spherical if $BRe \ll 10^2$. In the case of the bubble with radius 0.2 mm, $B \cong 6 \times 10^{-3}$ and $BRe \cong 0.5$.

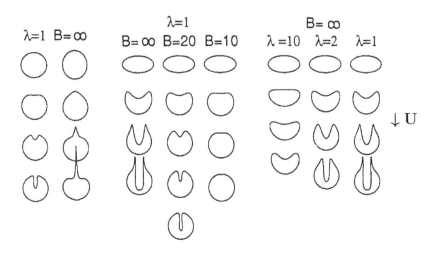

FIGURE 7.4
Numerical computation of the shape of a drop sedimenting downwards, as a function of the Bond number and the viscosity ratio. A slightly prolate drop develops a tail, while a slightly oblate drop develops a dimple. For $B = 10$, the surface tension is large enough to restore the spherical shape. A large internal viscosity does not prevent the instability from developing. (From Stone [49], reproduced with permission from the *Annual Reviews of Fluid Mechanics*.)

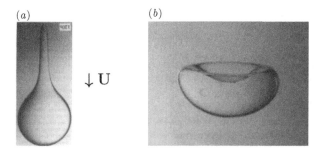

FIGURE 7.5
Experimental shape of a droplet sedimenting under large Bond number: (a) From Koh and Leal [34]; (b) From Baumann et al. [8]. (Reproduced with permission from the American Institute of Physics.)

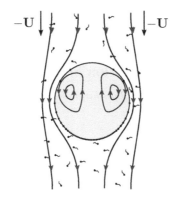

FIGURE 7.6
Marangoni effect on a droplet sedimenting with velocity **U** in a fluid at rest containing surfactant molecules. The accumulation of molecules in the downstream part of the interface creates a surface tension gradient which opposes the fluid motion along the interface.

Thus, the bubble remains spherical, as is observed for the bubbles in sparkling drinks.

For a liquid drop sedimenting in another immiscible liquid, the density difference is smaller, the viscosity is higher and the surface tension is smaller than in the preceding case. It is thus impossible to draw conclusions regarding the maximum size of the drop.

Finally, it happens frequently that surfactant molecules are present in the suspending liquid. A surfactant molecule has a polar hydrophilic head and an organic hydrophobic chain. Such molecules tend to adsorb at interfaces (air-water or oil-water). They are then swept by the motion towards the downstream part of the drop where they accumulate before desorbing. This non-uniform surfactant concentration creates a surface tension gradient and a surface traction that opposes the fluid motion. A spherical cap at the back of the bubble has thus a very low interfacial velocity. This in turns perturbs the global droplet motion as shown schematically in Figure 7.6. This phenomenon is called the Marangoni effect and occurs whenever a flow is perturbed or created by a surface tension gradient (see Problem 7.4.1).

7.4   Problems

7.4.1   Thermocapillary Motion of a Gas Bubble

From an Ecole Polytechnique problem written with K. Moffatt

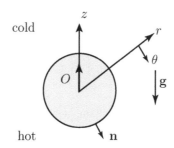

FIGURE 7.7
Gas bubble moving in a fluid under the influence of thermocapillary forces
due to a temperature gradient.

The objective of the problem is to study the motion of a gas bubble suspended
in a quiescent liquid subjected to a temperature gradient. The variation of
temperature creates a surface tension variation on the bubble interface which
can be large enough to change the direction of bubble motion.

A gas bubble with radius a is suspended in a Newtonian liquid (viscosity
μ, density ρ) in which there is a uniform gradient of temperature T

$$\frac{dT}{dz} = -\alpha, \quad (\alpha = \text{Cst}, \quad \alpha > 0)$$

where the z-axis is directed along the upward vertical direction (Figure 7.7).
The surface tension between the gas and the liquid varies linearly with temperature,

$$\frac{d\gamma_s}{dT} = -\beta, \quad (\beta = \text{Cst}, \quad \beta > 0)$$

in a given temperature range. We make the following assumptions:

- The only external force acting on the bubble is gravity.
- The bubble moves with velocity $U\mathbf{e}_z$ where U must be determined.
- The bubble deformation is neglected and the spherical shape is re-
  tained.
- The presence of the bubble does not perturb the temperature gra-
  dient to first order.

- The Reynolds number of the flow around the bubble is small enough for the Stokes equations to apply.

1. Use a spherical coordinate system with vertical z-axis, centred on the bubble and fixed with respect to the fluid at infinity. Write the flow boundary conditions in this coordinate system.

2. Determine the stream function Ψ.

3. Compute the force exerted by the fluid on the bubble.

4. Compute the bubble velocity and find the critical value α_c of the temperature gradient for which the bubble is motionless.

5. Analyse the normal stress condition and deduce whether the spherical shape is indeed a solution of the problem.

6. In the case where $U = 0$, draw a sketch of the streamlines in the fluid and explain how the flow is created.

7. Consider the case where the bubble moves up with velocity $U > 0$ and use a reference frame linked to the bubble. Find the stream function $\widetilde{\Psi}(r, \theta)$ in this reference frame. Determine the streamline corresponding to $\widetilde{\Psi}(r, \theta) = 0$ and show it on a small sketch.

7.4.2 Flow in a Droplet Due to an Electric Field

From an Ecole Polytechnique problem written with K. Moffatt

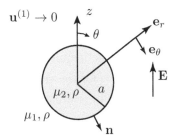

FIGURE 7.8
Electro-hydrodynamics of a liquid droplet suspended in another liquid at rest at infinity.

A droplet of a Newtonian liquid (2) (viscosity μ_2, density ρ) is suspended in another immiscible Newtonian liquid (1) (viscosity μ_1, same density ρ), as shown in Figure 7.8. The surface tension between the two liquids is denoted γ_s. The two liquids have different electrical conductivity and dielectric constant.

The liquid (1) is at rest far from the droplet. The drop is assumed to remain spherical with radius a.

We use a spherical coordinate system (r, θ, ϕ) centred on the drop. We apply a vertical uniform electric field $\mathbf{E} = E\mathbf{e}_z$ which perturbs the charge distribution on the surface of the drop. The resulting electrostatic effect is a force per unit surface area $\mathbf{T}$ which is exerted on the surface of the drop and given by

$$T_r = E^2(c + N\cos^2\theta), \quad T_\theta = E^2 S \sin\theta\cos\theta, \quad T_\phi = 0$$

where c, N and S have constant values which depend only on the electrical properties of the two fluids.

The interfacial stress $\mathbf{T}$ creates a flow $(\mathbf{u}^{(2)}, p^{(2)}, \boldsymbol{\sigma}^{(2)})$ inside the drop and $(\mathbf{u}^{(1)}, p^{(1)}, \boldsymbol{\sigma}^{(1)})$ in the suspending liquid. The objective of the problem is the study of these flows and the determination of the conditions for which the drop remains spherical.

The flow being obviously axisymmetric around the z-axis, we use the axisymmetric stream function $\Psi(r, \theta)$ which satisfies the bi-harmonic axisymmetric equation.

1. Show that the solutions to this Stokes problem are of the form

    $$\Psi_n(r, \theta) = r^n \sin 2\theta \cos\theta, \quad \text{where} \quad n = -2, 0, 3, 5$$

2. Let $\omega_n \mathbf{e}_\phi$ be the vorticity vector corresponding to the stream function Ψ_n. Compute ω_n for $n = -2, 0, 3, 5$.

3. Let p_n be the pressure field corresponding to the stream function Ψ_n. Compute the general expression of p_n for $n = -2, 0, 3, 5$ (assume that the flow occurs in a liquid with viscosity μ).

4. Show that the resultant force acting on the droplet is zero.

5. Write the boundary conditions for the velocity fields.

6. Write the force balance at the interface and deduce another boundary condition.

7. We seek a solution of the form

    $$\Psi = \begin{cases} \Psi^{(1)} = \left(A\frac{a^4}{r^2} + Ba^2\right)\sin^2\theta\cos\theta, & (r \geq a) \\[2mm] \Psi^{(2)} = \left(C\frac{r^3}{a} + D\frac{r^5}{a^3}\right)\sin^2\theta\cos\theta, & (r \leq a) \end{cases}$$

    where A, B, C and D are constant. Find the components of the velocities $u_r^{(1)}, u_\theta^{(1)}, u_r^{(2)}, u_\theta^{(2)}$.

8. Using the velocity boundary conditions, show that

    $$B = -A, \quad C = A, \quad D = -A$$

9. Sketch the streamlines in the two fluids.

10. Show that the stress components in liquids (1) and (2) on $r = a$ are given by

$$\sigma_{rr}^{(1)} = -p_0^{(1)} - 2\mu_1 \frac{A}{a}(3\cos^2\theta - 1), \qquad \sigma_{r\theta}^{(1)} = -10\mu_1 \frac{A}{a}\cos\theta\,\sin\theta,$$

$$\sigma_{rr}^{(2)} = -p_0^{(2)} + 3\mu_2 \frac{A}{a}(3\cos^2\theta - 1), \qquad \sigma_{r\theta}^{(2)} = 10\mu_2 \frac{A}{a}\cos\theta\,\sin\theta$$

where $p_0^{(1)}$ and $p_0^{(2)}$ are constant. Write the stress balance on the interface. Show that the droplet remains spherical only if the viscosity ratio is

$$\frac{\mu_1}{\mu_2} = \frac{9S - 10N}{10N - 6S}$$

Find the values of the ratio N/S for which it is possible to find an equilibrium state.

11. Discuss qualitatively what happens when μ_1/μ_2 is slightly different from the above value.

8

General Solutions of the Stokes Equations

CONTENTS

The solutions of the Stokes equations in Chapters 3 to 6 have been obtained in cases where the system geometry is simple enough (two-dimensional or axisymmetric) to allow us to find an analytical solution. In many real situations, the flow is three-dimensional and the system geometry is complex. We then seek numerical solutions or approximate analytical solutions by means of regular perturbation analysis (when possible). In all cases, the technique consists of using the linearity of the Stokes equations: we write the solution to the problem as a linear combination of fundamental solutions and determine the coefficients of the combination that allow us to satisfy the problem boundary conditions. Since the solution of the Stokes equations is unique, we are sure that the solution thus obtained is the one!

In this chapter we compute a new fundamental solution which corresponds to the flow created by a point force and we then derive an infinite series of other fundamental solutions. We show how the solution to a Stokes problem can be expressed as a series of fundamental solutions or as an integral equation which is solved numerically. The numerical modelling of Stokes flows is presently a very active domain of research with numerous applications in biophysics or industry (process engineering, microfluidics, biomedical engineering, environment, etc).

8.1   Flow Due to a Point Force

A simple fundamental solution to the Stokes equations is the flow due to a point force in a fluid at rest.

8.1.1   Stokeslet

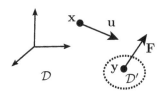

FIGURE 8.1
Velocity field $\mathbf{u}(\mathbf{x})$ due to a point force $\mathbf{F}$ located at $\mathbf{y}$.

A point force $\mathbf{F}$ is located at point $\mathbf{y}$ in a fluid domain $\mathcal{D}$ at rest at infinity. We seek the flow field at position $\mathbf{x}$ which is created by this force (Figure 8.1). The Stokes momentum equation is modified to account for $\mathbf{F}$:

$$\mu\nabla^2\mathbf{u} - \nabla p + \mathbf{F}\,\delta(\mathbf{x} - \mathbf{y}) = 0 \qquad (8.1)$$

where $\delta(\mathbf{x} - \mathbf{y})$ is a three-dimensional Dirac function such that

$$\delta(\mathbf{x} - \mathbf{y}) = 0 \qquad \text{for} \qquad \mathbf{x} \neq \mathbf{y}$$
$$\delta(\mathbf{x} - \mathbf{y}) \to \infty \qquad \text{for} \qquad \mathbf{x} = \mathbf{y}$$

$$\int_{\mathcal{D}'} \delta(\mathbf{x} - \mathbf{y})\mathbf{F}\,dV' = \mathbf{F} \qquad (8.2)$$

where $\mathcal{D}'$ is an arbitrary domain around $\mathbf{y}$.

We must of course add the continuity equation

$$\nabla \cdot \mathbf{u} = 0 \tag{8.3}$$

The solution of Equation (8.1) is in index notation

$$u_i(\mathbf{x}) = \frac{1}{8\pi\mu} G_{ik}(\mathbf{x} - \mathbf{y}) F_k \tag{8.4}$$

The tensor G_{ik} is the *stokeslet* or *Oseen–Burgers* tensor

$$G_{ik} = \frac{\delta_{ik}}{r_{xy}} + \frac{(x_i - y_i)(x_k - y_k)}{r_{xy}^3} \tag{8.5}$$

where $r_{xy} = |\mathbf{x} - \mathbf{y}|$. The pressure is given by (within an additional constant)

$$p = \frac{1}{8\pi} \Pi_k F_k \qquad \text{with} \qquad \Pi_k = \frac{2(x_k - y_k)}{r_{xy}^3} \tag{8.6}$$

The corresponding stress tensor is obtained from Newton's law:

$$\sigma_{ij} = \Sigma_{ijk} F_k \qquad \text{with} \qquad \Sigma_{ijk} = -\frac{3}{4\pi} \frac{(x_i - y_i)(x_j - y_j)(x_k - y_k)}{r_{xy}^5} \tag{8.7}$$

8.1.1.1 Demonstration

The Dirac function may be written as

$$\delta(\mathbf{x} - \mathbf{y}) = -\frac{1}{4\pi} \nabla^2 \left(\frac{1}{r_{xy}} \right) \tag{8.8}$$

The pressure p is a harmonic function (Chapter 2, Equation (2.6)) which is proportional to $\mathbf{F}$

$$p = \frac{-1}{4\pi} \mathbf{F} \cdot \nabla \left(\frac{1}{r_{xy}} \right) \tag{8.9}$$

where the factor $(-1/4\pi)$ is included to facilitate later computations. We replace p by expression (8.9) and use Equation (8.8) to find that Equation (8.1) becomes

$$\mu \nabla^2 u_i + \frac{1}{4\pi} F_k \left[\frac{\partial}{\partial x_k} \frac{\partial}{\partial x_i} \left(\frac{1}{r_{xy}} \right) - \delta_{ik} \nabla^2 \left(\frac{1}{r_{xy}} \right) \right] = 0 \tag{8.10}$$

The solution of Equation (8.10) which takes into account the incompressibility condition is then

$$u_i = \frac{1}{4\pi} F_k \left[\frac{\partial^2 H}{\partial x_k \partial x_i} - \delta_{ik} \nabla^2 H \right] \tag{8.11}$$

where H is an harmonic function such that

$$\nabla^2 H = \frac{-1}{\mu}\left(\frac{1}{r_{xy}}\right) \tag{8.12}$$

The solution is simply

$$H = -r_{xy}/2\mu \tag{8.13}$$

Replacing H by its expression (8.13) in (8.11), we find the fundamental solution (8.4), (8.5).

8.1.2 Stokeslet Properties

- The stokeslet (8.5) does indeed satisfy the continuity equation (8.3):

$$\frac{\partial G_{ik}}{\partial x_i} = \frac{-\delta_{ik}(x_i - y_i)}{r_{xy}^3} + \frac{\delta_{ii}(x_k - y_k) + \delta_{ik}(x_i - y_i)}{r_{xy}^3}$$
$$- 3\frac{(x_i - y_i)(x_i - y_i)(x_k - y_k)}{r_{xy}^5} = 0 \quad (8.14)$$

(recall that $\delta_{ii} = 3$).

- The force $\mathbf{F}'$ exerted on the boundary $\mathcal{S}'$ of a spherical domain $\mathcal{D}'$ with radius a, centred on $\mathbf{y}$, is equal to $-\mathbf{F}$ for any value of a. Indeed, $\mathbf{F}'$ is given by

$$F_i' = \int_{\mathcal{S}'} \sigma_{ij}\, n_j dS' \tag{8.15}$$

Using the expression of the stress tensor given by Equation (8.7) and noting that on a sphere $dS' = a^2 d\Omega$ where Ω is the solid angle, we find

$$F_i' = \frac{-3}{4\pi}F_k \int_{\mathcal{S}'} \frac{x_i x_j x_k}{a^3} n_j d\Omega \tag{8.16}$$

Noting that the outer unit normal vector to $\mathcal{S}'$ is $\mathbf{n} = \mathbf{x}/a$ and using the integration formula of Appendix A.3, we find

$$F_i' = \frac{-3}{4\pi}F_k \int_{\mathcal{S}'} n_i n_j n_k n_j d\Omega = \frac{-3}{4\pi}F_k \frac{4\pi}{3}\delta_{ik} = -F_i \tag{8.17}$$

We can also recover this result by writing Equation (8.1) in the form

$$\frac{\partial \sigma_{ij}}{\partial x_j} = -F_i\, \delta(x) \tag{8.18}$$

Applying Gauss' theorem to the integral of Equation (8.18) in $\mathcal{D}'$, we find the result

$$F_i' = \int_{\mathcal{S}'} \sigma_{ij}\, n_j\, dS = \int_{\mathcal{D}'} -F_i\, \delta(x)\, dV = -F_i \tag{8.19}$$

- The velocity field created by the point force decreases as $1/r_{xy}$ as we move away from $\mathbf{y}$:

$$G_{ik} \sim O(1/r_{xy}) \qquad \text{when} \qquad r_{xy} \to \infty \qquad (8.20)$$

- The solution (8.4) to (8.7) corresponds also to the velocity and pressure fields created at point $\mathbf{y}$ by a point force located at $\mathbf{x}$: we simply exchange the roles of $\mathbf{x}$ and $\mathbf{y}$. This operation does not modify $\mathbf{G}$ but changes the sign of Σ:

$$G_{ij}(\mathbf{x} - \mathbf{y}) = G_{ij}(\mathbf{y} - \mathbf{x}), \quad \Sigma_{ijk}(\mathbf{x} - \mathbf{y}) = -\Sigma_{ijk}(\mathbf{y} - \mathbf{x}) \qquad (8.21)$$

8.1.3 Correspondence with the Flow around a Sphere

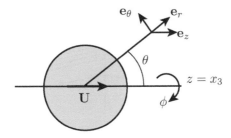

FIGURE 8.2
Flow around a solid sphere moving with velocity $\mathbf{U}$ in a quiescent fluid.

A stokeslet was first encountered when we studied the flow around a sphere of radius a, moving with translation velocity $\mathbf{U} = U\mathbf{e}_z$ in a fluid at rest (Chapter 6, Section 6.3). In a spherical coordinate system centred on the sphere and with z-axis colinear to $\mathbf{U}$ (Figure 8.2), the components of the flow field far from the sphere are given by Equation (6.53):

$$u_r = 2C\cos\theta/r, \quad u_\theta = -C\sin\theta/r, \quad C = 3Ua/4 \qquad (8.22)$$

where C is the stokeslet coefficient. Far away from a translating sphere, the velocity component in the x_3-direction (same as the z-direction of spherical coordinates, see Figure 8.2) is then

$$u_3 = u_r \cos\theta - u_\theta \sin\theta = C(1 + \cos^2\theta)/r \qquad (8.23)$$

The force exerted by the sphere on the fluid is given by the Stokes formula

$$\mathbf{F}_{\to\text{fluid}} = F_3\, \mathbf{e}_3 = 6\pi\mu a U \mathbf{e}_3 \qquad (8.24)$$

Replacing C by its value (8.22) and using Equation (8.24), we find

$$U = u_3 = \frac{F_3}{8\pi\mu}(1 + \cos^2\theta) \tag{8.25}$$

However, the flow field associated to a point force $\mathbf{F} = (0, 0, F_3)$ is given in the 3-direction by Equations (8.4) and (8.5):

$$u_3 = \frac{1}{8\pi\mu}G_{33}F_3 = \frac{F_3}{8\pi\mu}\left[\frac{1}{r} + \frac{x_3^2}{r^3}\right] = \frac{F_3}{8\pi\mu r}(1 + \cos^2\theta) \tag{8.26}$$

We thus see that the two definitions of the stokeslet are consistent. This means that 'seen from far away', the effect of a translating sphere in a fluid at rest is the same as the one created by a point force centred on the sphere (Figure 8.3).

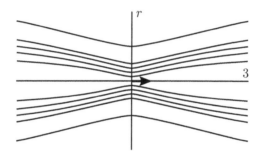

FIGURE 8.3
Streamlines of a stokeslet [0,0,F_3].

8.1.4   Solutions Derived from a Stokeslet

We now seek other fundamental solutions. We note first that if we have a fundamental solution, it is easy to generate an infinite series of others. Indeed, consider a velocity $\mathbf{u}(\mathbf{x})$ and pressure $p(\mathbf{x})$ that satisfy the Stokes equations

$$\mu\frac{\partial^2 u_i}{\partial x_k \partial x_k} - \frac{\partial p}{\partial x_i} = 0, \quad \frac{\partial u_i}{\partial x_i} = 0 \tag{8.27}$$

If we take the derivative of Equation (8.27) with respect to x_j and invert the order of derivation, we see immediately that $\partial u_i/\partial x_j$ and $\partial p/\partial x_j$ are also a solution of Equation (8.27).

We thus start from the velocity (8.4) and pressure (8.6) fields created by a point force located at the origin of coordinates ($\mathbf{y} = 0$). We take a linear combination of derivatives with respect to x_j and we obtain a new solution of the Stokes equation:

$$u_i = \frac{A_j}{8\pi\mu}\frac{\partial G_{ik}}{\partial x_j}F_k \tag{8.28}$$

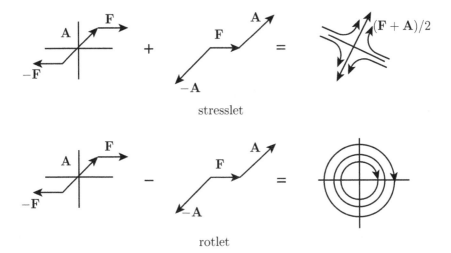

stresslet

rotlet

FIGURE 8.4
Flow field created by a force doublet centred on the origin. Decomposition into symmetric (stresslet) and antisymmetric (rotlet) parts.

where A_j are the constant coefficients of the linear combination. This velocity can be interpreted as the velocity field created by two equal and opposite forces centred near the origin (Figure 8.4). Indeed, consider the velocity created at point $\mathbf{x}$ by a point force $\mathbf{F}$ located at point $\mathbf{A}/2$, such that $|\mathbf{A}/2| \ll |\mathbf{x}|$. This velocity field can be expanded as a Taylor series:

$$u_i(\mathbf{x}) = \frac{1}{8\pi\mu} \left[G_{ik}(\mathbf{x} - 0)\, F_k + \frac{A_j}{2} \frac{\partial}{\partial x_j} G_{ik}(\mathbf{x} - 0)\, F_k + ... \right] \qquad (8.29)$$

Similarly, the velocity created at point $\mathbf{x}$ by a point force $-\mathbf{F}$ located at point $-\mathbf{A}/2$ is given by

$$u_i(\mathbf{x}) = \frac{1}{8\pi\mu} \left[G_{ik}(\mathbf{x} - 0)\, (-F_k) + \frac{-A_j}{2} \frac{\partial}{\partial x_j} G_{ik}(\mathbf{x} - 0)\, (-F_k) + ... \right]$$
$$(8.30)$$

When we add Equations (8.29) and (8.30), we obtain Equation (8.28). Thus the flow field (8.28) is due to a force doublet centred on the origin. The graphical representation of this mathematical operation is represented on the left of Figure 8.4.

The product $F_k A_j$ is a second-order tensor that we can chose with zero trace without loss of generality, since the trace is always multiplied by zero as shown in Equation (8.14). We can decompose $F_k A_j$ into the sum of a symmetric part S_{kj} and of an antisymmetric part $\varepsilon_{jmk}\gamma_m$ (Figure 8.4):

$$F_k A_j / 8\pi\mu = S'_{kj} + \varepsilon_{jmk}\gamma'_m \qquad \text{with} \qquad S'_{jj} = 0 \qquad (8.31)$$

where

$$S'_{kj} = (F_k A_j + F_j A_k)/8\pi\mu \qquad \text{and} \qquad \varepsilon_{jmk}\gamma'_m = (F_k A_j - F_j A_k)/8\pi\mu$$

After computing $\partial G_{ik}/\partial x_j$, we find

$$u_i = S_{lm}\frac{x_l x_m x_i}{r^5} + \varepsilon_{imk}\gamma_m\frac{x_k}{r^3} \qquad (8.32)$$

where $r = (x_i x_i)^{1/2}$, $S_{lm} = -3S'_{lm}$ and $\gamma_m = 2\gamma'_m$. The term S_{ij}, called the *stresslet*, corresponds to a straining flow centred on the origin (Figure 8.4). The pressure field due to the stresslet is

$$p = p_0 + 2\mu S_{lm}\frac{x_l x_m}{r^5} \qquad (8.33)$$

The antisymmetric term $\varepsilon_{imk}\gamma_m$, called the *rotlet*, corresponds to the velocity field created by a singular torque centred on the origin (Figure 8.4). The associated pressure field is constant.

If we derive again Equation (8.28) with respect to x_m, we obtain a *quadruplet*, etc.

8.2   Irrotational Solutions

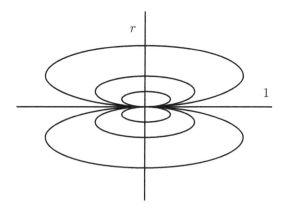

FIGURE 8.5
Streamlines of a potential dipole $(T_1, 0, 0)$.

Another class of flows that satisfies the Stokes equations corresponds to irrotational flows with thus a constant pressure field (Equation 2.14). The simplest

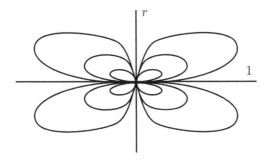

FIGURE 8.6
Streamlines of a potential quadrupole $T_{11} = -2T_{22}$, $T_{22} = T_{33}$, $T_{ij} = 0$ for $i \neq j$.

one is the flow created by a point source with intensity $4\pi T_0$ centred on the origin:

$$u_i = T_0 \frac{x_i}{r^3}, \qquad p = p_0 \qquad (8.34)$$

Following the procedure of Section 8.1.4, we take the derivative and obtain a new solution: the potential dipole (Figure 8.5)

$$u_i = -\frac{T_i}{r^3} + 3\frac{T_l x_l x_i}{r^5}, \qquad p = p_0 \qquad (8.35)$$

The derivative of the potential dipole yields the potential quadrupole (Figure 8.6)

$$u_i = 6\frac{T_{im} x_m}{r^5} - 15\frac{T_{lm} x_l x_m x_i}{r^7}, \qquad p = p_0, \quad T_{jj} = 0 \qquad (8.36)$$

8.3 Series of Fundamental Solutions: Singularity Method

As we have seen with the stokeslet, it is possible to obtain an infinite series of solutions to the Stokes equations by taking the successive derivatives of a given solution. These particular solutions are called 'singularities'.

The singularity method consists of seeking the linear combination of singularities that allows us to satisfy the problem boundary conditions. The reasoning is similar to the one which is used to solve two-dimensional potential flow problems where the solution is sought by superposing elementary flows (source, sink, dipole, etc). However, since the analytical complexity of the singularities increases significantly with the order of the derivatives, we are quickly limited. In practice, we use only low-order fundamental solutions.

If this is not possible, we turn towards a numerical solution as explained in Section 8.4.

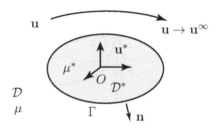

FIGURE 8.7
Liquid droplet suspended in another immiscible liquid subjected to the flow $\mathbf{u}^\infty$ far from the drop.

A typical problem consists of computing the flow around a liquid drop (domain $\mathcal{D}^*$) freely suspended in another immiscible liquid (domain $\mathcal{D}$) subjected to flow $\mathbf{u}^\infty$ far from the droplet (Figure 8.7). The problem equations are given in Chapter 7, Section 7.1.1. This problem was solved for the simple case of a drop translating in a fluid at rest in Chapter 7, Section 7.2, where we took advantage of the problem symmetry to use cylindrical coordinates. The singularity method allows us to tackle this problem for more general flow situations. We define a Cartesian coordinate system centred inside the drop (for example, on the centre of mass and then seek a solution to the problem of the form

$$\mathbf{u} = \mathbf{u}^\infty + \sum \mathbf{u}_s \quad \text{for} \quad \mathbf{x} \in \mathcal{D} \tag{8.37}$$

$$\mathbf{u}^* = \sum \mathbf{u}_s^* \quad \text{for} \quad \mathbf{x} \in \mathcal{D}^* \tag{8.38}$$

where $\sum \mathbf{u}_s$ and $\sum \mathbf{u}_s^*$ represent linear combinations of singularities. In order to satisfy the condition at infinity (7.3), we must use for $\mathbf{u}_s$ singularities that go to zero as r goes to infinity (e.g. a stokeslet, a stresslet, etc). However, these singularities are singular when r goes to zero. It is clear that they do not qualify for the internal flow $\mathbf{u}^*$ since they are singular in the domain $\mathcal{D}^*$. The internal flow field $\mathbf{u}^*$ must then be expanded in terms of other singularities that are regular at $r = 0$.

We now show how those singularities can be systematically generated.

8.3.1   Singularities for External Flows

In the preceding sections, we have computed singularities which were of order 2 at most. Higher-order singularities are easily generated. We first note that the pressure satisfies the Laplace equation (Chapter 2, Section 2.1.1) and can thus be expressed as a sum of harmonic fundamental functions:

$$p = \sum_{n=-\infty}^{-1} P_n \qquad (8.39)$$

with

$$P_{-(m+1)} = S_{ijk...} \frac{(-1)^m}{1.3.5...(2m-1)} \frac{\partial^m \left(\frac{1}{r}\right)}{\partial x_i \partial x_j \partial x_k....} \qquad m = 0, 1, 2, ... \quad (8.40)$$

where the tensors $S_{ijk...}$ are constant, symmetric with respect to any permutation of their indices and have a zero contraction with respect to any couple of indices. For example, the first terms of the series are

$$p = p_0 + \frac{S_0}{r} + S_i \frac{x_i}{r^3} + S_{ij} \frac{x_i x_j}{r^5} + ... \qquad (8.41)$$

One can verify that expressions (2.61) and (8.41) are indeed equivalent when the zero trace condition is imposed on S_{ij}. These functions decrease when r increases and are singular for $r = 0$. The general expression for the velocity field follows from Equations (2.7) to (2.9) and is given by

$$u_i = \sum_{n=-\infty}^{-1} \varepsilon_{ijk} \frac{\partial R_n}{\partial x_j} x_k + \frac{\partial \Phi_n}{\partial x_i} + \frac{(n+3)r^2}{2(n+1)(2n+3)} \frac{\partial P_n}{\partial x_i} - \frac{n}{(n+1)(2n+3)} x_i P_n$$

$$(8.42)$$

where Φ_n and R_n are also harmonic functions with expressions identical to that of P_n, except for different values of the tensor coefficients. The pressure field associated to Φ_n and R_n is constant. The expression (8.42) is a linear combination of singularities initiated from the stokeslet (P_n and R_n) and from irrotational solutions (Φ_n). The first terms of the series correspond to the singularities computed in Section 8.1.4 and Section 8.2.

The selection of which singularities to use depends on the boundary conditions and on the type of flow. For example, the stokeslet and the potential dipole are associated with a constant velocity of the flow, as seen for a solid sphere (Chapter 6, Section 6.3) or for a gas bubble (Chapter 7, Section 7.2). The stresslet and the potential quadrupole are associated with a linear shear flow in the suspending fluid. The rotlet is associated with a vorticity field. If we were to study a particle suspended in a parabolic flow field (e.g. Poiseuille flow), we would have to use higher-order singularities (of order 3 at least) since a quadratic flow is described with order 2 and 3 tensors.

8.3.2 Singularities for Internal Flows

When we consider an internal problem (i.e. such that the origin of the coordinate system is inside the domain, e.g. $\mathcal{D}^*$), we seek fundamental solutions that

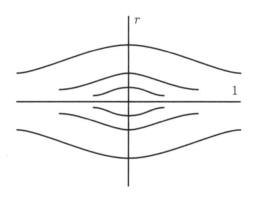

FIGURE 8.8
Streamlines of a stokeson $(A_1, 0, 0)$.

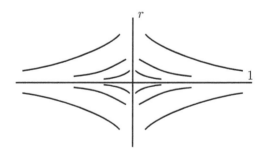

FIGURE 8.9
Streamlines of an axisymmetric internal quadrupole $A_{11} = -A_{22}$, $A_{22} = A_{33}$, $A_{ij} = 0$ for $i \neq j$.

are regular for $r = 0$. We thus start with solutions of the Laplace equation that are regular for $r = 0$:

$$p = \sum_{n=1}^{\infty} P_n \qquad (8.43)$$

with

$$P_n = A_{ijk\ldots} \frac{(-1)^n r^{2n+1}}{1.3.5\ldots(2n-1)} \frac{\partial^n \left(\frac{1}{r}\right)}{\partial x_i \partial x_j \partial x_k \ldots}, \qquad n = 1, 2, \ldots \qquad (8.44)$$

where the tensors $A_{ijk\ldots}$ are constant with the same properties as the tensors $S_{ijk\ldots}$. The first terms of the series are thus

$$p = p_0 + A_i x_i + A_{ij} x_i x_j + \ldots \qquad (8.45)$$

The velocity field is given by Equation (8.42), where the sum is now taken over positive values of n between 1 and ∞. We thus obtain the first low-order singularities

- A uniform flow (Φ_1)

$$u_i = B_i, \quad p = p_0 \qquad (8.46)$$

- A *roton* (R_1)

$$u_i = \varepsilon_{ijk} C_j x_k, \quad p = p_0 \qquad (8.47)$$

- A *stokeson* (P_1)

$$u_i = 2r^2 A_i - A_m x_m x_i, \quad p = p_0 + 10\mu A_m x_m \qquad (8.48)$$

- An *internal quadrupole* (P_2)

$$u_i = 5r^2 A_{im} x_m - 2A_{lm} x_l x_m x_i, \quad p = p_0 + 21\,\mu A_{lm} x_l x_m, \quad A_{mm} = 0 \qquad (8.49)$$

- A *stresson* (Φ_2)

$$u_i = B_{im} x_m, \quad p = p_0, \quad B_{mm} = 0 \qquad (8.50)$$

The velocity fields of a stokeson and of an internal quadrupole are shown in Figures 8.8 and 8.9, respectively. The stresson corresponds to a linear shear flow (e.g. for a simple shear flow in the $x_1 x_2$-plane, the stresson components are $B_{12} = B_{21} \neq 0$ with all other components equal to zero).

For an internal problem, the choice of singularities is guided by the boundary conditions and by the flow in the external fluid. For example, the uniform flow and the stokeson are associated with a uniform translation motion of the external liquid, whereas the stresson and the internal quadrupole are associated with an external linear shear flow.

The velocity, pressure and stress fields of singularities of order 2 at most are given in Tables 8.1 and 8.2.

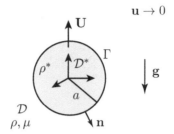

FIGURE 8.10
Gas bubble moving with velocity **U** in a fluid at rest. A Cartesian coordinate
system is used.

8.3.3   Example: Translating Gas Bubble

We turn again to the problem of a gas bubble translating with velocity **U**
in a fluid at rest (Chapter 7, Section 7.2). The advantage of the singularity
method is that we can use a simple Cartesian coordinate system instead of
spherical coordinates. We again neglect the velocity field inside the bubble
and assume that the external flow is governed by the Stokes equations. The
boundary conditions are then (Chapter 7, Section 7.2)

$$\mathbf{u} \to 0 \quad \text{for} \quad |\mathbf{x}| \to \infty \tag{8.51}$$

and for $\mathbf{x} \in \Gamma$

$$u_i n_i = U_i n_i \tag{8.52}$$

$$\sigma_{ik} n_k - (\sigma_{km} n_k n_m) n_i = 0 \tag{8.53}$$

$$\sigma'_{km} n_k n_m - (-p'^*) = \gamma \kappa \tag{8.54}$$

Since the far field flow is a uniform flow (in a reference system linked to
the bubble), we write the external flow in domain $\mathcal{D}$ as the sum of a stokeslet
and a potential dipole (Table 8.2):

$$u_i = \frac{S_i}{r} + \frac{S_m x_m x_i}{r^3} - \frac{T_i}{r^3} + \frac{3T_m x_m x_i}{r^5} \tag{8.55}$$

The associated stress traction is (Table 8.2)

$$\sigma_{ij} n_j = -p_0 n_i + 2\mu \left[\frac{-3S_m x_m x_i}{r^4} + \frac{3T_i}{r^4} - \frac{9T_m x_m x_i}{r^6} \right] \tag{8.56}$$

The tensors **S** and **T** are determined from the boundary conditions on the bubble surface since the condition at infinity (8.51) is already satisfied by Equation (8.55). We first compute the velocity component normal to the spherical
bubble surface

$$u_i n_i = \frac{S_i}{a} \frac{x_i}{a} + \frac{S_m x_m x_i}{a^3} \frac{x_i}{a} - \frac{T_i}{a^3} \frac{x_i}{a} + \frac{3T_m x_m x_i}{a^5} \frac{x_i}{a} \tag{8.57}$$

where we have used the fact that the unit normal vector to the sphere $r = a$ is given by

$$n_i = x_i/a \tag{8.58}$$

Using $x_i x_i = a^2$ on the sphere surface, we can simplify Equation (8.57)

$$u_i n_i = 2\frac{S_i x_i}{a^2} + \frac{2T_i x_i}{a^4} \tag{8.59}$$

Using boundary condition (8.52), we obtain a first relation

$$2\frac{S_i x_i}{a^2} + \frac{2T_i x_i}{a^4} = U_i \frac{x_i}{a} \tag{8.60}$$

or

$$2\mathbf{S}/a + 2\mathbf{T}/a^3 = \mathbf{U} \tag{8.61}$$

The tangential stress component is easily computed from Equation (8.56) and condition (8.53) becomes

$$\sigma_{ij} n_j - \sigma_{lm} n_l n_m n_i = 6\mu \left[\frac{T_i}{a^4} - \frac{T_m x_m x_i}{a^6} \right] = 0 \tag{8.62}$$

We thus obtain a second relation,

$$\mathbf{T} = 0 \tag{8.63}$$

The solution of Equations (8.56) and (8.63) is then

$$\mathbf{T} = 0 \quad \text{and} \quad \mathbf{S} = \mathbf{U}a/2 \tag{8.64}$$

The effect of the bubble is a stokeslet only. The drag force on the bubble is then the stokeslet coefficient (Section 8.1.2 and Table 8.2)

$$\mathbf{F} = -8\pi\mu\mathbf{S} = -4\pi\mu a\mathbf{U} \tag{8.65}$$

We note that this solution of the problem is much easier and faster to obtain than the one based on the use of spherical coordinates and a stream function.

In the case of a liquid droplet, it is not possible to neglect the internal flow. Inside the drop, we write the flow as the sum of regular singularities. The boundary conditions at the interface then allow us to determine the coefficients of all the singularities (see Problem 8.5.1.)

8.3.4 Applications of the Singularity Method

The first computation of the break-up a liquid droplet in a linear shear flow [1, 24], were based on the singularity method and a perturbation expansion to compute the flow inside and around the drop as well as its deformation. The same technique was also used to determine the deformation of liquid droplets encapsulated by an elastic membrane [2, 3] or of elastic spheres representing macromolecules [25]. The advantage of these studies is that they provide *analytical* relations between the suspending flow parameters, the physical properties of the system and the particle motion and deformation. The singularity method is limited to particles whose shape remains nearly spherical. Typically it cannot be used to model the large deformation of low-viscosity drops ($\lambda \ll 1$) which tend to become highly elongated with pointed cusps at the end. The other extreme is the case of slender bodies [4] or of ellipsoidal particles [15, 16] for which the singularity method can be used. In these cases, the singularities are distributed on the axis of the particle and we seek the distribution density that allows us to satisfy the no-slip condition on the surface of the body. One must realise that the computations quickly become quite cumbersome and are limited to regular shapes.

The series representation of the Stokes solution can also be used to optimise numerical solutions of Stokes flows. For example, when a solid or deformable particle is moving near a wall, it is interesting to introduce a solution to the Stokes equations that satisfies identically the no-slip condition on the wall. This is done by adding singularities by the image reflexion method [11, 58]. The same technique can also be used for problems involving interactions between flowing particles. For more details, see the book by Kim and Karrila [32].

8.4   Integral Form of the Stokes Equations

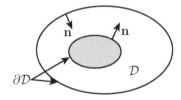

FIGURE 8.11
Definition of the fluid domain.

When the flow geometry is complex, it is difficult to obtain an analytical solution and we thus have to find a numerical solution to the problem. If we wish to know the pressure and velocity at each point of the domain $\mathcal{D}$, the classical numerical techniques developed for the Navier–Stokes equations can be used (finite differences, finite elements, volume of fluid, etc). However, it may happen that we are interested in the flow in some parts of $\mathcal{D}$ only or that we only want to determine the velocity of the domain boundary and the forces acting on it. If such is the case, it is not necessary to compute the flow everywhere in $\mathcal{D}$. The boundary integral method is then very efficient as it is based on the determination of the force and velocity distribution on the domain boundary. A further advantage of the method is that it allows us to include moving and/or deformable boundaries (liquid interfaces, elastic walls, etc) without having to mesh again the whole flow domain.

8.4.1 Velocity Field in $\mathcal{D}$

We consider a fluid domain $\mathcal{D}$ bounded by a piecewise continuous boundary $\partial \mathcal{D}$. The unit normal vector to $\partial \mathcal{D}$ pointing inside $\mathcal{D}$ is denoted $\mathbf{n}$ (Figure 8.11). We seek the solution $(\mathbf{u}, \boldsymbol{\sigma})$ of the Stokes equations in $\mathcal{D}$:

$$\nabla \cdot \mathbf{u} = 0 \quad \text{and} \quad \nabla \cdot \boldsymbol{\sigma} = 0 \tag{8.66}$$

with mixed boundary conditions

$$\mathbf{u} = \mathbf{U}(\mathbf{x}) \quad \text{for } \mathbf{x} \in \partial \mathcal{D}_1 \tag{8.67}$$

$$\boldsymbol{\sigma} \cdot \mathbf{n} = \mathbf{f}^B(\mathbf{x}) \quad \text{for } \mathbf{x} \in \partial \mathcal{D}_2 \tag{8.68}$$

where $\partial \mathcal{D} = \partial \mathcal{D}_1 \cup \partial \mathcal{D}_2$ and where $\mathbf{U}(\mathbf{x})$ and $\mathbf{f}^B(\mathbf{x})$ are given.

We use the integral form (2.51) of the general reciprocal theorem. Flow (1) is the real flow $(\mathbf{u}, \boldsymbol{\sigma})$ that we seek to determine. Flow (2) is the stokeslet flow created by a point force $\mathbf{F}$ located at $\mathbf{x}$ (Equations 8.4 to 8.7). The two flows occur in the same liquid $(\mu^{(1)} = \mu^{(2)} = \mu)$, so that the viscosity can be eliminated from Equation (2.51). The reciprocal theorem then becomes

$$\int_{\partial \mathcal{D}} u_i(\mathbf{y}) \Sigma_{ijk}(\mathbf{y} - \mathbf{x}) F_k [-n_j(\mathbf{y})] dS_y$$

$$- \frac{1}{8\pi\mu} \int_{\partial \mathcal{D}} G_{ik}(\mathbf{y} - \mathbf{x}) F_k \sigma_{ij}(\mathbf{y}) [-n_j(\mathbf{y})] dS_y = \int_{\mathcal{D}} \left[u_i(\mathbf{y}) \frac{\partial \sigma_{ij}^{(2)}}{\partial y_j} \right] dV_y \tag{8.69}$$

where the notation dS_y and dV_y for the surface and volume elements indicates that integration is taken over $\mathbf{y}$. The minus sign in front of the normal vector is due to the fact that the reciprocal theorem uses a unit normal vector pointing out of $\mathcal{D}$. The right-hand side of Equation (8.69) is computed from Equation (8.18)

$$\int_{\mathcal{D}} \left[u_i \frac{\partial \sigma_{ij}^{(2)}}{\partial y_j} \right] dV_y = \int_{\mathcal{D}} -u_i(\mathbf{y}) \delta(\mathbf{y} - \mathbf{x}) F_i \, dV_y$$

$$= -u_k(\mathbf{x}) F_k \quad \text{if} \quad \mathbf{x} \in \mathcal{D}$$

$$= 0 \quad \text{if} \quad \mathbf{x} \notin \mathcal{D} \tag{8.70}$$

We can now eliminate F_k from Equations (8.69) and (8.70) to obtain an integral form of the Stokes equations

$$-\frac{1}{8\pi\mu}\int_{\partial D}G_{ik}(\mathbf{y}-\mathbf{x})f_i(\mathbf{y})dS_y + \int_{\partial D}u_i(\mathbf{y})\Sigma_{ijk}(\mathbf{y}-\mathbf{x})n_jdS_y$$

$$= u_k(\mathbf{x}) \qquad \mathbf{x}\in D$$
$$= 0 \qquad \mathbf{x}\notin D \quad (8.71)$$

where $\mathbf{f} = \boldsymbol{\sigma}\cdot\mathbf{n}$ is the force per unit surface area exerted by D on ∂D. The equation (8.71) relates the velocity at any point of the domain D to the traction and velocity distribution on the boundary ∂D. The first integral (with $\mathbf{G}$) is called the single layer potential while the second integral (with $\boldsymbol{\Sigma}$) is called the double layer potential (by analogy with the electrostatic fields created by a charge or dipole distribution on a surface). The tensors G_{ik} and Σ_{ijk} are the Green functions of a Stokes flow in an infinite domain.

8.4.2   Velocity Field on the Boundary ∂D

The values of $\mathbf{u}$ and $\mathbf{f}$ on ∂D are computed from Equation (8.71) and from the boundary conditions (8.67) and (8.68). However, one must be careful when applying the boundary conditions at a point $\mathbf{x}$ on the boundary, because the Green functions are singular when $\mathbf{y} = \mathbf{x}$ (see definitions (8.5) and (8.7)). We must then take the limit of Equation (8.71) when $\mathbf{x}\to\partial D$. The single layer integral (with $\mathbf{G}$) is continuous when $\mathbf{x}\to\partial D$. The double layer integral (with $\boldsymbol{\Sigma}$) is discontinuous when $\mathbf{x}\to\partial D$:

$$\lim_{x\to\partial D}\int_{\partial D}u_i(\mathbf{y})\Sigma_{ijk}(\mathbf{y}-\mathbf{x})n_jdS_y = \pm\frac{1}{2}u_k(\mathbf{x}) + \int_{\partial D}^{PV}u_i(\mathbf{y})\Sigma_{ijk}(\mathbf{y}-\mathbf{x})n_jdS_y$$
$$(8.72)$$

where the plus sign corresponds to the case where $\mathbf{x}$ tends to ∂D from inside D. The minus sign corresponds to the case where $\mathbf{x}$ tends to ∂D from outside D. The integral on the right-hand side is the principal value integral taken over $\{\partial D - \mathbf{x}\}$. The limit exists only if the normal vector to ∂D is continuous and thus if the surface is locally planar (∂D is a Lyapounov surface).

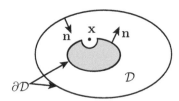

FIGURE 8.12
Integration for $\mathbf{x}\in\partial D$: we avoid the singularity $\mathbf{x} = \mathbf{y}$ by integrating over a half sphere S_ε with centre $\mathbf{x}$ and radius $\varepsilon\to 0$.

8.4.2.1 Demonstration

For $\mathbf{x} \in \partial\mathcal{D}$, we take the integration over $\partial\mathcal{D}$ less half a sphere S_ε centred on $\mathbf{x}$ with radius ε. For example in Figure 8.12, $\mathbf{x}$ tends to $\partial\mathcal{D}$ from inside $\mathcal{D}$ ($\mathbf{x} \to \partial\mathcal{D}^+$), so that $\mathbf{y} - \mathbf{x} = -\varepsilon\mathbf{n}$. We then take the limit $\varepsilon \to 0$:

$$\lim_{\varepsilon \to 0} \int_{S_\varepsilon} G_{jk}(\mathbf{y} - \mathbf{x}) f_j(\mathbf{y}) dS_y = f_j(\mathbf{x}) \lim_{\varepsilon \to 0} \int_{S_\varepsilon} \left[\frac{\delta_{jk}}{\varepsilon} + \frac{n_j n_k}{\varepsilon} \right] \varepsilon^2 d\Omega = 0 \quad (8.73)$$

and

$$\begin{aligned}
\lim_{\varepsilon \to 0} \int_{S_\varepsilon} u_i(\mathbf{y}) \Sigma_{ijk}(\mathbf{y} - \mathbf{x}) n_j dS_y &= u_i(\mathbf{x}) \lim_{\varepsilon \to 0} \int_{S_\varepsilon} \left[-\frac{3}{4\pi} \frac{(-n_i n_j n_k)}{\varepsilon^2} \right] n_j \varepsilon^2 d\Omega \\
&= u_i(\mathbf{x}) \left(\frac{3}{4\pi} \right) \frac{1}{2} \frac{4\pi}{3} \delta_{ik} = \frac{1}{2} u_k(\mathbf{x}) \quad (8.74)
\end{aligned}$$

where $d\Omega$ is the solid angle element centred on $\mathbf{x}$. The $1/2$ factor comes from the fact that we integrate over half a sphere. The limit as taken here exists only if the surface appears to be locally plane (thus the restriction to Lyapounov surfaces). The rest of the integral on $\partial\mathcal{D}$ corresponds the the principal value.

8.4.3 Boundary Integral Method

In conclusion, the boundary integral form of the Stokes equations is

$$\alpha\, u_k(\mathbf{x}) = -\frac{1}{8\pi\mu} \int_{\partial\mathcal{D}} G_{jk}(\mathbf{y} - \mathbf{x}) f_j(\mathbf{y}) dS_y + \int_{\partial\mathcal{D}}^{PV} u_i(\mathbf{y}) \Sigma_{ijk}(\mathbf{y} - \mathbf{x}) n_j dS_y \quad (8.75)$$

with

- $\alpha = 0$ if $\mathbf{x} \notin \mathcal{D}$
- $\alpha = 1$ if $\mathbf{x} \in \mathcal{D}$
- $\alpha = 1/2$ if $\mathbf{x} \in \partial\mathcal{D}$ $(\mathbf{x} \to \partial\mathcal{D}^+)$

The boundary integral method is based on the use of the boundary conditions to determine the velocity and force distribution on the domain boundaries. For example, if the velocity $\mathbf{U}(\mathbf{x})$ is given on all the boundary $\partial\mathcal{D}$, the force distribution $\mathbf{f}(\mathbf{x})$ is obtained from equation

$$-\frac{1}{8\pi\mu} \int_{\partial\mathcal{D}} G_{jk}(\mathbf{y} - \mathbf{x}) f_j(\mathbf{y}) dS_y = \frac{1}{2} U_k(\mathbf{x}) - \int_{\partial\mathcal{D}}^{PV} U_i(\mathbf{y}) \Sigma_{ijk}(\mathbf{y} - \mathbf{x}) n_j dS_y \quad (8.76)$$

which is a Fredholm integral equation of the first kind. Similarly, if the force $\mathbf{f}^B(\mathbf{x})$ is given on all the boundary $\partial\mathcal{D}$, the velocity distribution $\mathbf{U}(\mathbf{x})$ is obtained from equation

$$\frac{1}{2} U_k(\mathbf{x}) - \int_{\partial\mathcal{D}}^{PV} U_i(\mathbf{y}) \Sigma_{ijk}(\mathbf{y} - \mathbf{x}) n_j dS_y = \frac{-1}{8\pi\mu} \int_{\partial\mathcal{D}} G_{jk}(\mathbf{y} - \mathbf{x}) f_j^B(\mathbf{y}) dS_y \quad (8.77)$$

which is a Fredholm integral equation of the second kind. In both cases, those integrals are easy to solve numerically. For more complex boundary conditions such as Equations (8.67) and (8.68), the problem is more difficult. For further information, see the book by Pozrikidis [44].

8.4.4   Motion of a Solid Particle

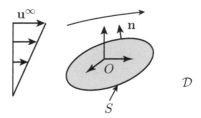

FIGURE 8.13
Flow around a solid particle.

We can use the boundary integral technique to compute the flow around a solid particle suspended in a liquid which is subjected to a Stokes flow $\mathbf{u}^\infty$ far from the particle. The boundary conditions are then

$$\mathbf{u} \to \mathbf{u}^\infty(\mathbf{x}) \quad \text{for} \quad \mathbf{x} \to \infty \qquad (8.78)$$

and

$$\mathbf{u}(\mathbf{x}) = \mathbf{u}^{(S)}(\mathbf{x}) = \mathbf{U} + \boldsymbol{\Omega} \times \mathbf{x} \quad \text{for} \quad \mathbf{x} \in V \qquad (8.79)$$

where V represents the interior of the particle and $\mathbf{x}$ is measured from a point O inside the particle (e.g. the centre of mass) as shown in Figure 8.13. The integral equation (8.75) can then be simplified to

$$u_k(\mathbf{x}) \;=\; u_k^\infty(\mathbf{x}) \;-\; \frac{1}{8\pi\mu} \int_S G_{jk}(\mathbf{y} - \mathbf{x}) f_j(\mathbf{y}) dS_y \qquad (8.80)$$

where S is the surface of the particle and $\mathbf{x}$ a material point in $\mathcal{D}$ or on S (for the demonstration, see Problem 8.5.4). This relation is more general than the mobility equations introduced in Chapter 6, because it allows to account for a non-zero far fluid motion (recall that in Chapter 6, the fluid is assumed to be at rest far from the particle). When we write Equation (8.80) for $\mathbf{x}$ on S, we can compute

- Either the force distribution (or the resultant) that must be applied on the particle to obtain a given motion

- Or the particle motion resulting from the combined effect of the hydrodynamic forces due to the flow $\mathbf{u}^\infty$ and of other external forces and torques acting on the particle.

8.4.5 Applications of the Boundary Integral Method

The first application of Equation (8.80) was proposed by Youngren and Acrivos [56], who demonstrated the feasibility of the boundary integral method to obtain a numerical solution to Stokes flow problems. Since that time, the boundary integral method has been extensively used to solve a number of problems. To cite but a few, it has been used to compute the flow near a rough surface, in two-dimensional corners or cavities. Indeed, the corner eddies shown in Figure 3.13 have been computed with this method [27]. The integral method is particularly well adapted to problems with free surfaces (liquid droplets, biological cells, capsules). In this type of problem, the free surface is deformed by the viscous traction $\mathbf{f}$ exerted by the flow. The position and geometry of part of $\partial\mathcal{D}$ is then unknown. However, some boundary conditions are imposed on this free surface (no slip condition for velocities and balance between viscous and interfacial forces). This type of problem is solved by successive approximations: we start with a given configuration of the free surface, solve Equation (8.75) and determine the velocity and stress distribution on the free surface. This allows us to move to the new configuration of the free surface and so on until a convergence criterion is satisfied. The motion and deformation of different particles have thus been studied, such as

- Liquid droplets freely suspended in a shear flow [31] or in a capillary tube [40]
- Capsules freely suspended in a shear flow [35, 39, 45, 48, 54, 57] or in a capillary tube [20, 37, 38].

These references are just cited as examples, as the research domain is very active and numerous studies are published every year with improved numerical techniques to solve the integral formulation of the Stokes equations.

TABLE 8.1
Singularities for Internal Flows

Uniform field Φ_1 associated with uniform velocity $\mathbf{U}$

$$u_i = B_i, \qquad p = p_0, \qquad \sigma_{ik} = -p_0\delta_{ik}$$

Stokeson P_1 associated with uniform velocity $\mathbf{U}$

$$u_i = 2r^2 A_i - A_m x_m x_i, \quad p = p_0 + 10\mu A_m x_m$$

$$\sigma_{ik} = -p_0\delta_{ik} + 3\mu(-4\delta_{ik}A_m x_m + A_i x_k + A_k x_i)$$

Internal quadrupole P_2 associated with a linear shear flow $\mathbf{e}$

$$u_i = 5r^2 A_{il}x_l - 2x_i A_{lm}x_l x_m, \qquad p = p_0 + 21A_{lm}x_l x_m$$

$$\sigma_{ik} = -p_0\delta_{ik} + \mu\left[-25A_{lm}x_l x_m\delta_{ik} + 10r^2 A_{ik} + 6(A_{im}x_k x_m + A_{km}x_i x_m)\right]$$

Roton R_1 associated with a torque $\mathbf{G}$ and/or a rotation velocity $\boldsymbol{\Omega}$

$$u_i = \varepsilon_{ijk}C_j x_k, \qquad p = p_0$$

$$\sigma_{ik} = -p_0\delta_{ik}$$

Stresson Φ_2 associated with a linear shear flow $\mathbf{e}$

$$u_i = 2B_{il}x_l, \qquad p = p_0$$

$$\sigma_{ik} = -p_0\delta_{ik} + 4\mu B_{ik}$$

TABLE 8.2
Singularities for External Flows

Stokeslet P_{-2} associated with a force $\mathbf{F}$ and/or a velocity $\mathbf{U}$

$$u_i = S_l \left[\frac{\delta_{il}}{r} + \frac{x_i x_l}{r^3} \right], \quad p = p_0 + 2\mu \frac{x_l}{r^3} S_l, \quad \sigma_{ik} = -p_0 \delta_{ik} - 6\mu \frac{x_i x_k x_l}{r^5} S_l$$

$$S_i = F_i / 8\pi\mu$$

Stresslet P_{-3} associated with a linear shear flow $\mathbf{e}$

$$u_i = S_{lm} \frac{x_l x_m x_i}{r^5}, \quad p = p_0 + 2\mu S_{lm} \frac{x_l x_m}{r^5}$$

$$\sigma_{ik} = -p_0 \delta_{ik} + \mu S_{lm} \left[\frac{1}{r^5} (\delta_{il} x_k x_m + \delta_{im} x_k x_l + \delta_{km} x_i x_l + \delta_{kl} x_i x_m) - \frac{10 x_i x_k x_l x_m}{r^7} \right]$$

Rotlet R_{-2} associated with a torque $\mathbf{G}$ and/or a rotation velocity $\mathbf{\Omega}$

$$u_i = \varepsilon_{imk} \gamma_m \frac{x_k}{r^3}, \quad p = p_0, \quad \sigma_{ik} = -p_0 \delta_{ik} - 3\mu\gamma_m \frac{1}{r^5} \left[\varepsilon_{iml} x_l x_k + \varepsilon_{kml} x_l x_i \right]$$

Source/sink Φ_{-1}

$$u_i = T_0 \frac{x_i}{r^3}, \quad p = p_0 \quad \sigma_{ik} = -p_0 \delta_{ik} + \mu T_0 \left[\frac{2\delta_{ik}}{r^3} - \frac{6 x_i x_k}{r^5} \right]$$

Potential dipole Φ_{-2} associated with a velocity $\mathbf{U}$

$$u_i = -\frac{T_i}{r^3} + 3 \frac{T_l x_l x_i}{r^5}, \quad p = p_0$$

$$\sigma_{ik} = -p_0 \delta_{ik} + \mu T_l \left[\frac{6}{r^5} (\delta_{ik} x_l + \delta_{il} x_k + \delta_{kl} x_i) - \frac{30 x_i x_k x_l}{r^7} \right]$$

Potential quadrupole Φ_{-3} associated with a linear shear flow $\mathbf{e}$

$$u_i = 6 \frac{T_{im} x_m}{r^5} - 15 \frac{T_{lm} x_l x_m x_i}{r^7}, \quad p = p_0$$

$$\sigma_{ik} = -p_0 \delta_{ik} + \mu \left[12 \frac{T_{ik}}{r^5} - \frac{30}{r^7} (2 T_{il} x_l x_k + 2 T_{kl} x_l x_i + \delta_{ik} T_{lm} x_l x_m) + \frac{210 T_{lm} x_i x_k x_l x_m}{r^9} \right]$$

8.5 Problems

8.5.1 Liquid Droplet Translating in a Quiescent Liquid

A spherical liquid drop (viscosity μ^*, density ρ^*, radius a) is moving with constant velocity $\mathbf{U}$ in a fluid at rest at infinity (viscosity μ, density ρ). Use the singularity method to compute the flow field. Compute the drag on the drop and rederive the results of Hadamard–Rybczynski (Chapter 7, Section 7.3.1).

8.5.2 Solid Sphere Freely Suspended in a Linear Shear Flow

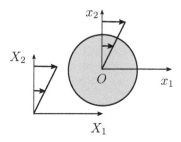

FIGURE 8.14
Solid sphere freely suspended in a simple shear flow. The reference frame $[Ox_1, x_2, x_3]$ is centred on the sphere and moves at time t with the undisturbed fluid velocity at the sphere centre.

A solid sphere (radius a, density ρ) is freely suspended in a viscous liquid (viscosity μ, density ρ). Far from the sphere, the suspending liquid is subjected to the unperturbed linear shear flow

$$u_i^\infty = e_{ik} X_k, \quad p = p_0$$

in the laboratory reference frame (X_1, X_2, X_3). The shear rate tensor e_{ij} is constant. There is no external force acting on the sphere and the flow Reynolds number is very small (usually because the sphere is small). The objective of the problem is to compute the velocity u_i and pressure p in the fluid. We introduce perturbation fields defined by

$$u_i' = u_i - u_i^\infty \quad \text{and} \quad p' = p - p_0$$

1. Write the flow field at infinity in the reference frame (x_1, x_2, x_3) moving with the fluid velocity at the sphere centre and centred on the sphere (Figure 8.14).

2. Give the boundary conditions for the perturbation fields.

3. Show that we can write the perturbation fields as the sum of a stresslet and of a quadrupole, and find the coefficients of these two singularities.

4. We consider a general linear shear flow

$$u_i^\infty = e_{ik}X_k + \omega_{ik}X_k$$

where the vorticity tensor ω_{ik} is antisymmetric ($\omega_{ik} = -\omega_{ki}$). Show that the perturbation velocity field is given by the same expression as in the previous question and that the sphere has a solid body rotation with angular velocity Ω_i given by $\varepsilon_{ijk}\Omega_j = \omega_{ik}$.

8.5.3 Hydrodynamic Interaction between Three Spheres

From an Ecole Polytechnique problem written with F. Dias

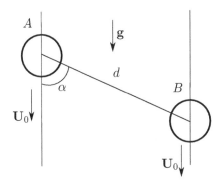

FIGURE 8.15
Sedimentation of two identical spheres under the effect of gravity $\mathbf{g} = -g\mathbf{e}_z$.

The objective of the problem is to study the influence of a third particle located on the path of a particle pair which falls in a Newtonian viscous liquid (viscosity μ, density ρ) under the the influence of gravity. The fluid is at rest at infinity. The flow field is three-dimensional. The three particles are identical homogeneous solid spheres with density ρ_s ($\rho_s > \rho$) and radius a.

1. The stream function of the flow due to one sphere moving with velocity $\mathbf{U}$ in a fluid at rest far from the sphere is given in spherical coordinates by (Chapter 6, Section 6.3)

$$\psi(r,\theta) = \frac{1}{4}Ua^2\left(\frac{3r}{a} - \frac{a}{r}\right)\sin^2\theta$$

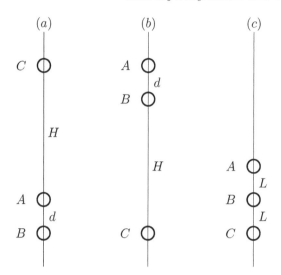

FIGURE 8.16
Sedimentation of three particles: (a) sphere C is located on top of the pair (A, B); (b) sphere C is located under the pair (A, B); (c) three equidistant spheres.

where $U = |\mathbf{U}|$. The coordinate system is centred on the sphere and the axis $\theta = 0$ is parallel to $\mathbf{U}$. The stream function is the sum of a stokeslet and of a potential dipole. We neglect the contribution of the potential dipole to the flow when it is less than 1% of the stokeslet contribution. At which distance D from the sphere centre is the dipole contribution negligible?

2. Find the sphere velocity $\mathbf{U} = -U\mathbf{e}_z$ when it moves under the effect of gravity $\mathbf{g} = -g\mathbf{e}_z$ only.

3. We consider now the sedimentation of two spheres A and B when the distance d between their centres is such that $d \gg a$ and $d > D$. The configuration is shown in Figure 8.15 in the vertical plane containing the two spheres centres. We note α, the angle between the centre line and the vertical direction. Each sphere moves with velocity $\mathbf{U}$

$$\mathbf{U} = \mathbf{U_0} + \frac{a}{d}\mathbf{U_1} + \cdots$$

where the neglected terms are $O((a/d)^2)$. What is the value of $\mathbf{U_0}$?

4. Show that at the next order, each sphere creates a velocity $(a/d)\,\mathbf{U_1}$ on the other sphere. Give the expression of $\mathbf{U_1}$.

5. Show that the distance d and the angle α do not change during the motion of the two spheres.

6. Show that the pair of spheres trajectory makes an angle γ with the vertical direction. Compute γ in terms of a, d and α. Find the condition for which the trajectory of the sphere centres remains vertical.

7. When the two spheres are one on top of the other ($\alpha = 0$), show that $\mathbf{U_1} = -\frac{3}{2}U\,\mathbf{e}_z$.

8. We consider a pair of spheres A and B, on the same vertical line and separated by d ($d \gg a$ and $d > D$). This pair falls with velocity $\mathbf{U_0} + (a/d)\,\mathbf{U_1}$. A third sphere C with radius a is located above the pair, on the same vertical line at a distance $H \gg d$ (Figure 8.16a). Compute the velocity of each sphere. Describe the motion of the three spheres and the configuration evolution in time.

9. The sphere C is located under the pair AB, on the same vertical line at a distance $H \gg d$ (Figure 8.16b). Describe the motion of the three spheres and the configuration evolution in time.

10. The three spheres are on the same vertical line and are separated by the same distance L ($L \gg a$) (Figure 8.16c). Describe the motion of the three spheres and the configuration evolution in time.

8.5.4 Integral Equation for the Flow around a Solid Particle

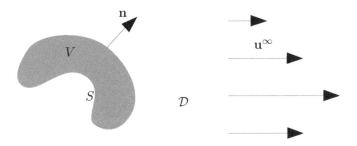

FIGURE 8.17
Rigid particle in an arbitrary Stokes flow.

The objective of this problem is to obtain Equation (8.80) which gives the velocity $u_k(\mathbf{x})$ around a solid particle suspended in a viscous fluid subjected to a Stokes flow $\mathbf{u}^\infty$ far from the particle. Let V and S denote the inside volume and the surface of the particle, respectively. The fluid domain $\mathcal{D}$ is bounded by S and by a surface S_∞ located far from the particle (Figure 8.17). The velocity of a particle material point is $\mathbf{u}^{(S)} = \mathbf{U} + \mathbf{\Omega} \times \mathbf{x}$. We introduce the perturbation velocity and stress fields

$$\mathbf{u}' = \mathbf{u} - \mathbf{u}^\infty, \qquad \boldsymbol{\sigma}' = \boldsymbol{\sigma} - \boldsymbol{\sigma}^\infty$$

1. Give the boundary conditions for the perturbation velocity.

2. Apply the integral formulation to the perturbation velocity $\mathbf{u}'$.

3. Apply an integral formulation to the solid body velocity field $\mathbf{u}^{(S)} - \mathbf{u}^{\infty}$.

4. Combine the two formulations to obtain Equation (8.80).

8.5.5   Integral Equation for the Flow around a Liquid Drop

The objective of the problem is to find the integral form for the flow around a liquid droplet (Chapter 7, Section 7.1.1). A liquid drop (viscosity $\mu^* = \lambda\mu$ and density ρ) is freely suspended in another non-miscible liquid (viscosity μ and density ρ) subjected to a Stokes flow $\mathbf{u}^{\infty}$ far from the drop. We note $\mathcal{D}^*$ and $\mathcal{D}$ the fluid domains corresponding to the drop and to the suspending liquid, respectively, while S is the separating interface. Both the internal and external flows satisfy the Stokes equations.

Write the integral form of the Stokes equations in each fluid domain and find the velocity of the points on S.

9

Introduction to Suspension Mechanics

CONTENTS

In the preceding chapters, we have studied the motion of a *single* particle in a flowing fluid. In practice, the particle being seldom alone in the flow, is subjected to hydrodynamic interactions with its neighbours. When the particle concentration increases, these collective effects become important and may result in non-linear properties of the bulk suspension, which may be very different from the initial properties of the suspending liquid. For example, mayonnaise has a viscosity which is different from that of its components (oil, vinegar or egg yolk).

The objective of suspension mechanics is to relate the motion of the microscopic suspended particles to the global macroscopic behaviour of the 'homogenised' fluid. The research on suspensions started in the early 20th century (Einstein's PhD thesis) and swiftly expanded at the beginning of the 1970s when Batchelor [5, 6] established a rigorous framework for the study of any suspension. In the 1970s and 1980s, the Cambridge (UK) and Stanford (USA) schools produced a number of renown scientists who have used Batchelor's framework to analyse various suspensions which were not amenable to rigorous analysis until then.

Presently, suspension mechanics is a very active research domain both on fundamental aspects and on applications. It is possible to distinguish between two major classes of problems depending on the role of volume forces (e.g. gravity) which act on the particles. When volume forces are important, they lead to an inhomogeneous repartition of particles. This phenomenon may be either desired (separation by means of sedimentation or centrifugation) or avoided (stabilisation of suspensions by means of physiochemical effects). In the other class of problems, the volume forces are negligible with respect to the viscous forces exerted by the flow. The problem then consists of determining the constitutive law of the suspension assuming that the spatial particle repartition is uniform.

The objective of this chapter is to present the main concepts of suspension mechanics and to show how they can be applied to analyse fairly complex situations.

9.1   Homogenisation of a Suspension

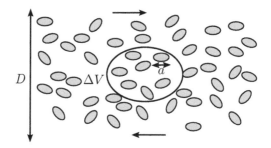

FIGURE 9.1
The control volume ΔV must contain many particles (characteristic dimension a) and be small with respect to the flow scale D.

Particles with characteristic dimension a, volume V_p, surface S_p and density ρ^* are suspended in a Newtonian incompressible liquid with density ρ and viscosity μ (Figure 9.1). The system of fluid plus particles is set into motion, so that there is at each point a local velocity $\mathbf{u}(\mathbf{x}, t)$ and a rate of strain tensor $\mathbf{e}(\mathbf{x}, t)$ with characteristic dimension denoted $\dot{\gamma}$ (unit: s^{-1}). The volume concentration c is the relative volume of the particles in a volume ΔV of suspension

$$c = \frac{\text{volume of particles} \ \in \ \Delta V}{\Delta V} \tag{9.1}$$

In this chapter, we consider the case where gravity forces $(\rho - \rho^*)\,\mathbf{g}$ are negligible with respect to the hydrodynamic forces exerted by the flow. We thus assume that the spatial repartition of particles is uniform. For identical particles, we can relate c to the number n_p of particles per unit volume:

$$c = n_p V_p \qquad (9.2)$$

The suspension will be treated as a homogenous liquid if it is possible to define an elementary volume ΔV with characteristic dimension $\Delta V^{1/3}$ such that

- ΔV contains enough particles to be statistically representative of the suspension

- $\Delta V^{1/3}$ is small compared to the scale D of variation of the flow: this means that ΔV can be considered as a 'point' volume at this scale

These two conditions are expressed as follows:

$$a \ll \Delta V^{1/3} \ll D \qquad (9.3)$$

We can then define the macroscopic values of the stress and of the rate of strain tensor of the suspension:

$$< \boldsymbol{\sigma} > = \frac{1}{\Delta V} \int_{\Delta V} \boldsymbol{\sigma}(\mathbf{x}) \, dV \qquad (9.4)$$

$$< \mathbf{e} > = \frac{1}{\Delta V} \int_{\Delta V} \mathbf{e}(\mathbf{x}) \, dV \qquad (9.5)$$

where macroscopic quantities are in brackets. For a simple shear flow in the 12-plane, the apparent shear viscosity μ_a of the suspension is defined as

$$\mu_a = \frac{< \sigma >_{12}}{2 < e >_{12}} \qquad (9.6)$$

Note that the homogenisation hypothesis implies that the particles must be small with respect to the flow scale. This also implies that the flow Reynolds number seen by a particle is usually very small $Re = \rho a^2 \dot{\gamma}/\mu \ll 1$. Consequently, the flow around a particle will be governed by the Stokes equations.

9.2   Micro–Macro Relationship

Batchelor [5, 6] has developed the theory to obtain constitutive laws for two-phase media with a random microscopic structure. The objective is to find a constitutive law

$$< \boldsymbol{\sigma} > = \mathcal{G}(< \mathbf{e} >, t, c, ...)$$

where $\boldsymbol{\mathcal{G}}$ is a tensor function to be determined and where the dots represent the relevant physical properties of the particles. We thus seek to evaluate the mean stress in the suspension, keeping in mind that ΔV contains both fluid (domain V_f) and particles (domain ΣV_p):

$$< \boldsymbol{\sigma} >= \frac{1}{\Delta V} \int_{\Delta V} \boldsymbol{\sigma}\, dV = \frac{1}{\Delta V} \int_{\Sigma V_p} \boldsymbol{\sigma}\, dV + \frac{1}{\Delta V} \int_{V_f} \boldsymbol{\sigma}\, dV \qquad (9.7)$$

The integral on the particles can be written as

$$\int_{\Sigma V_p} \sigma_{ij}\, dV = \int_{\Sigma V_p} \frac{\partial(\sigma_{ik} x_j)}{\partial x_k}\, dV - \int_{\Sigma V_p} \frac{\partial \sigma_{ik}}{\partial x_k} x_j\, dV \qquad (9.8)$$

Using Gauss' theorem for the first integral of the right-hand side and noting that $\nabla \cdot \boldsymbol{\sigma} + \mathbf{f} = 0$, where $\mathbf{f}$ is the external force per unit volume acting on the particles, we find

$$\int_{\Sigma V_p} \sigma_{ij}\, dV = \int_{\Sigma S_p} \sigma_{ik} x_j n_k\, dS + \int_{\Sigma V_p} f_i x_j\, dV \qquad (9.9)$$

The last term of the righ-hand side of Equation (9.9) is the resultant external force on the particles which plays an important role during sedimentation. In the rest of this chapter, this term will be assumed to be negligible (i.e. $\rho^* \cong \rho$).

To evaluate the integral over the fluid domain, we first introduce Newton's constitutive law

$$\int_{V_f} \sigma_{ij}\, dV = \int_{V_f} (-p\delta_{ij} + 2\mu e_{ij})dV \qquad (9.10)$$

and then decompose the fluid domain $V_f = \Delta V - \Sigma V_p$

$$\int_{V_f} \sigma_{ij}\, dV = \int_{V_f} -p\delta_{ij}\, dV + \int_{\Delta V} 2\mu e_{ij}\, dV - \int_{\Sigma V_p} 2\mu e_{ij}\, dV \qquad (9.11)$$

Replacing e_{ij} by its definition and using again Gauss' theorem, we can replace the integral over V_p by a surface integral taken over the surface S_p of the particles

$$\int_{V_f} \sigma_{ij}\, dV = \int_{V_f} -p\delta_{ij}\, dV + \int_{\Delta V} 2\mu e_{ij}\, dV - \int_{\Sigma S_p} \mu(u_i n_j + u_j n_i)\, dS \quad (9.12)$$

and using Equation (9.5) we obtain

$$\int_{V_f} \sigma_{ij}\, dV = \int_{V_f} -p\delta_{ij}\, dV + 2\mu\Delta V < e_{ij} > - \int_{\Sigma S_p} \mu(u_i n_j + u_j n_i)\, dS$$
$$(9.13)$$

Finally, the bulk stress in the suspension can be written as

$$< \sigma_{ij} >= -P\delta_{ij} + 2\mu < e_{ij} > + \Sigma^p_{ij} \qquad (9.14)$$

where the particle contribution Σ_{ij}^p to the stress tensor is given by

$$\Sigma_{ij}^p = \frac{1}{\Delta V}\Sigma \int_{S_p} \frac{1}{2}(\sigma_{ik}n_k x_j + \sigma_{jk}n_k x_i - \frac{2}{3}\delta_{ij}\sigma_{mk}n_k x_m)dS$$

$$-\frac{1}{\Delta V}\Sigma \int_{S_p} \mu(u_i n_j + u_j n_i)dS \quad (9.15)$$

The sums are taken over all the particles in ΔV. In Equation (9.15), we have used the symmetric stress deviator of $\sigma_{ik}n_k x_j$, which leads us to modify the isotropic term $P\delta_{ij}$. This operation has no incidence when the fluid and the particles are incompressible, since the pressure is then defined within an arbitrary isotropic constant. The contribution to the integral of the term $\mu(u_i n_j + u_j n_i)$ is identically zero for a particle undergoing solid body motion $u_i = U_i + \varepsilon_{ijk}\Omega_j x_k$.

It is also possible to write the velocity and stress fields under the form

$$u_i = u_i^\infty + u_i^p, \quad \sigma_{ij} = \sigma_{ij}^\infty + \sigma_{ij}^p \quad (9.16)$$

where u_i^∞ and σ_{ij}^∞ are the bulk velocity and stress fields imposed on the suspending fluid and where u_i^p and σ_{ij}^p are the perturbed fields due to the presence of the particles. Equation (9.15) becomes

$$\Sigma_{ij}^p = \frac{1}{\Delta V}\Sigma \int_S \frac{1}{2}(\sigma_{ik}^p n_k x_j + \sigma_{jk}^p n_k x_i - \frac{2}{3}\delta_{ij}\sigma_{mk}^p n_k x_m)dS$$

$$-\frac{1}{\Delta V}\Sigma \int_S \mu(u_i^p n_j + u_j^p n_i)dS \quad (9.17)$$

where the integrals are taken over a surface S surrounding a particle without crossing another one. Note that even if the particle is solid, the contribution of $\mu(u_i^p n_j + u_j^p n_i)$ is not zero because it involves the *perturbation* of velocity.

It is clear that in order to evaluate the contribution of the particles to the bulk stress, we must first determine the flow field around each particle and then compute the righ-hand side of Equations (9.15) or (9.17). This task represents a formidable hydrodynamics problem for which an exact solution can be obtained in some particular simple cases only. Indeed, the flow field around a particle depends not only on its physical properties (size, geometry, deformability, etc), but also on the position, motion and physical properties of the neighbouring particles. The computation of Equation (9.15) is thus very difficult to perform in general.

However, in the case of *identical solid particles*, Equation (9.15) can be simplified:

$$\Sigma_{ij}^p =$$

$$n_p \int \left\{ \int_{S_p} \frac{1}{2}\left[\sigma_{ik}n_k x_j + \sigma_{jk}n_k x_i - \frac{2}{3}\delta_{ij}\sigma_{mk}n_k x_m\right]dS \right\} P(\mathbf{B}, \mathcal{C}\,|O)d\mathbf{B}\,d\mathcal{C}$$

$$(9.18)$$

where:

- The integration is performed over a typical particle
- **B** is a second order symmetric tensor that defines the orientation of the solid particle (e.g. **B** could be the inertia tensor of the particle). In the case of an axisymmetric particle, the tensor **B** can be replaced by the unit vector **b** oriented along the revolution axis
- $\mathcal{C}$ represents the suspension configuration, that is, the position, orientation and velocity of all the particles in the suspension (or more modestly, in the statistically representative volume ΔV)
- $P(\mathbf{B}, \mathcal{C}|O)$ is the conditional probability of finding the typical particle centred on O, with orientation **B** and the rest of the suspension in the configuration $\mathcal{C}$

In conclusion, in order to compute the stress in a suspension, we must find the velocity field in the fluid and determine the configuration of all the particles. The two problems are obviously closely linked.

9.3   Dilute Suspension

A simple situation occurs when the suspension is very dilute ($c \ll 1$). In this case, the distance between two particles is large enough for the hydrodynamic interactions to become asymptotically negligible. Note that this can be readily achieved since the perturbation flow decreases as $1/r^2$, where r is the distance measured from one particle, when there is no body force, that is, no stokeslet (see Problem 8.5.2). In a dilute suspension, each particle then behaves as if it were alone in the flow and the configuration is meaningless in Equation (9.18). The particle stress in a dilute suspension of solid particles thus becomes

$$\Sigma_{ij}^p = n_p \int \left\{ \int_{S_p} \frac{1}{2} \left[\sigma_{ik} n_k x_j + \sigma_{jk} n_k x_i - \frac{2}{3} \delta_{ij} \sigma_{mk} n_k x_m \right] dS \right\} P(\mathbf{B}) d\mathbf{B}$$

(9.19)

9.3.1   Dilute Suspension of Identical Spheres

When the suspended particles are identical solid spheres with radius a, the orientation **B** is irrelevant (!) and the particle stress becomes simply

$$\Sigma_{ij}^p = n_p \int_{S_p} \frac{1}{2} \left[\sigma_{ik} n_k x_j + \sigma_{jk} n_k x_i - \frac{2}{3} \delta_{ij} \sigma_{mk} n_k x_m \right] dS$$

(9.20)

where the integration is taken over the surface S_p of a typical sphere. We can also use Equation (9.17) to write the particle stress in the form

$$\Sigma_{ij}^p =$$
$$\frac{1}{\Delta V}\Sigma \int_{S_\infty}\left[\frac{1}{2}(\sigma_{ik}^p n_k x_j + \sigma_{jk}^p n_k x_i - \frac{2}{3}\delta_{ij}\sigma_{mk}^p n_k x_m) - \mu(u_i^p n_j + u_j^p n_i)\right]dS$$
(9.21)

where the integration is taken over a sphere S_∞, centred on the sphere and with radius $r_\infty \gg a$. The surface of S_∞ will not cross any other particle since the suspension is dilute. The perturbation flow around a sphere freely suspended in a shear flow $\mathbf{u}^\infty = \mathbf{e}^\infty \cdot \mathbf{x} + \boldsymbol{\omega}^\infty \cdot \mathbf{x}$ is the sum of a stresslet and of a potential quadrupole (Chapter 8, Problem 8.5.2):

$$u_i^p = \frac{S_{km}x_k x_m x_i}{r^5} + \frac{6T_{ik}x_k}{r^5} - \frac{15T_{km}x_k x_m x_i}{r^7}$$

with

$$S_{ij} = -5a^3 e_{ij}^\infty/2, \quad T_{ij} = -a^5 e_{ij}^\infty/6$$

The stresslet alone contributes to the integral (9.21). After a straightforward calculation, we find

$$\Sigma_{ij}^p = -n_p\frac{8}{3}\pi\mu S_{ij}$$
(9.22)

Replacing S_{ij} by its value and introducing the volume concentration c, we obtain

$$\Sigma_{ij}^p = n_p\frac{20}{3}\pi a^3\mu e_{ij}^\infty = 5c\mu e_{ij}^\infty = 5c\mu < e_{ij} >$$
(9.23)

where the average value of the shear rate is equal to the the value at infinity (dilution hypothesis). The bulk stress in the suspension then becomes

$$< \sigma_{ij} > = -P\delta_{ij} + 2\mu(1 + \frac{5}{2}c) < e_{ij} >$$
(9.24)

A dilute suspension of spherical particles behaves as a Newtonian liquid with apparent viscosity μ_a larger than the viscosity μ of the suspending liquid

$$\frac{\mu_a}{\mu} = 1 + \frac{5}{2}c$$
(9.25)

This is the famous Einstein formula, which was obtained as part of Einstein's PhD thesis [21]. In practice, this formula is valid for monodisperse suspensions of spheres with a concentration less than 3% (see Problem 9.7.1).

9.3.2 Dilute Suspension of Anisotropic Particles

The geometrically simplest non-spherical particle is the ellipsoid. The flow field around an ellipsoidal particle freely suspended in a linear shear flow has been computed by Jeffery [30]. An axisymmetric ellipsoid (also called a spheroid) is very interesting because it can represent approximately a wide variety of anisotropic particles from fibres to disks, depending on the value of the axis ratio. The suspension is dilute if

$$c = n_p L^3 \ll 1$$

where L is the longest axis of the ellipsoid. The direction of the revolution axis is defined by the unit vector $\mathbf{b}$. When a spheroid is suspended in a viscous liquid subjected to the general linear shear flow with constant shear rate e_{ik}^∞ and vorticity ω_{ik}^∞

$$u_i^\infty = e_{ik}^\infty x_k + \omega_{ik}^\infty x_k$$

it takes a periodic rotational motion with an angular velocity which varies in time and space

$$\partial b_i / \partial t = \omega_{ik}^\infty b_k + f[e_{ik}^\infty b_k - b_i(e_{mk}^\infty b_m b_k)] \tag{9.26}$$

where f is a complicated function of the axis ratio. When it is alone in the fluid, a spheroid will remain indefinitely on its orbit. The velocity and stress on the ellipsoid surface depend on $\mathbf{b}$ and are thus periodic functions of time. Consequently, in order to compute the integral (9.19), we must first determine the probability density of orientation $P(\mathbf{b})$, an impossible task in general because the orientation of each particle at time t depends on its initial orientation at time t_0. There are a few situations though where the computation of $P(\mathbf{b})$ is possible:

(a) Pure straining motion with no vorticity (entrance of a filter or a converging die, extrusion)

(b) Particles all initially aligned (e.g. by a magnetic field for appropriate particles)

(c) Strong Brownian motion

In case (a), the particles are aligned by the flow which allows us to compute $P(\mathbf{b})$. The computation has been performed for fibres and disks by Hinch and Leal [28].

In case (b), the theory predicts that the viscosity of the suspension is a periodic function of time. Experiments have been performed on a dilute suspension of identical fibres all aligned at time $t = 0$ [43]. It was found that the suspension apparent viscosity was indeed a periodic function of time at the start of flow, but that the amplitude of the viscosity oscillations decreased quickly with time and that the apparent viscosity eventually became constant. This indicates that the suspension loses the memory of its initial state. This behaviour is probably due to different phenomena that were neglected in the theoretical analysis: far-field interaction effects (small but finite), not perfectly identical fibres, small inertia effects.

In case (c), Brownian motion creates an isotropic statistical repartition of the ellipsoids. If the Brownian motion is strong compared to the shear flow, the statistical orientation repartition is isotropic and Equation (9.19) can be computed [28]. Of course this situation is encountered only for very small particles (e.g. DNA strands) subjected to moderate shear flows.

In conclusion, we note that it is difficult if not impossible to obtain a general $O(c)$ expression for the viscosity of a dilute suspension of anisotropic particles.

9.3.3 Approximation $O(c^2)$ to the Viscosity of a Suspension of Spheres

Since the validity of the Einstein formula is limited to a very small concentration, it is interesting to seek the next term $O(c^2)$ of the series expansion of the viscosity in terms of the volume concentration. This means that we now take into account the hydrodynamic interactions between the spheres that we had previously neglected. The main contribution to the particle stress Σ_{ij}^p remains that of a stresslet which must be modified to account for the presence of neighbouring particles. The series expansion of Equation (9.22) is then

$$\Sigma_{ij}^p = n_p S_{ij}^{(0)} + n_p \int \Delta S_{ij}(r) P(r \,|O)dS \qquad (9.27)$$

where $\mathbf{S}^{(0)}$ is the stresslet of an isolated sphere, $\Delta \mathbf{S}$ is the stresslet variation due to the pair interaction with a neighbour sphere and $P(r|O)$ is the conditional probability of finding a sphere at distance r from the test sphere (Figure 9.2). A first difficulty arises because the integral in Equation (9.27) is not absolutely convergent! Indeed, $\Delta \mathbf{S}$ is $O(r^{-3})$, $P(r|O)$ is $O(1)$ and dS is $O(r^2)$. The second difficulty is linked to the computation of $P(r|O)$ (also called the pair density probability).

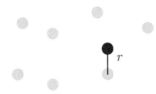

FIGURE 9.2
Pair interaction in a semi-dilute suspension of identical spheres.

Indeed, the motion of two spheres freely suspended in a simple shear flow is well-known, and it thus possible to compute $\Delta \mathbf{S}$ and the relative trajectories of the two spheres (which of course depend on their initial relative positions) [7]. The probability density $P(r|O)$ can then be determined from the possible trajectories, provided these are open. Thus the computation of $P(r|O)$ is possible for a pure straining motion with no vorticity such that all trajectories extend from $-\infty$ to $+\infty$. After solving the convergence problem of the integral in Equation(9.27), Batchelor and Green [7] found the apparent viscosity of a suspension of spheres subjected to an elongation flow:

$$\mu_a/\mu = 1 + 2.5c + 7.6c^2 \qquad (9.28)$$

In the case of a simple shear flow (with vorticity), the relative trajectories can be either open or closed. This last situation occurs when the two spheres form a doublet that rotates in the flow. It is then impossible to compute the general form of the $O(c^2)$ term of the apparent shear viscosity of a suspension of identical spheres subjected to simple shear flow.

However, when the spheres are small enough ($a < 1\ \mu$m) to be subjected to Brownian motion, it is possible to compute this $O(c^2)$ term under certain conditions. Brownian motion creates a random motion of the spheres which is characterised by a diffusion coefficient D_0, given for a sphere by

$$D_0 = kT/6\pi\mu a$$

where k is the Boltzmann constant and T the temperature (°K). The Péclet number Pe measures the ratio between the the Brownian diffusion time and the viscous diffusion time imposed by the flow

$$Pe = \dot{\gamma}\, a^2/D_0 = \mu\dot{\gamma}\, a^3/kT \qquad (9.29)$$

where $\dot{\gamma}$ is the shear rate magnitude and where the 6π factor is ignored. When $Pe \ll 1$, Brownian diffusion is the dominant phenomenon which leads to a uniform repartition of spheres. As a consequence, the pair probability density is uniform and isotropic. In this case and for any linear shear flow, the apparent viscosity of the suspension is

$$\mu_a/\mu = 1 + 2.5c + 6.1c^2 \qquad (9.30)$$

The term $6.1\,c^2$ includes the additional dissipation due to perturbations of the Brownian motion by the flow. Indeed, if we compute the apparent viscosity of a semi-dilute suspension of spheres without Brownian motion but with an isotropic repartition of spheres (due to an unspecified phenomenon), we find an apparent shear viscosity which is slightly lower than the value (9.30)

$$\mu_a/\mu = 1 + 2.5c + 5.2c^2$$

Those computations show how difficult it is to obtain an analytical expression for the apparent viscosity a suspension, even for a suspension of identical spheres which seemed a priori to be a most simple case.

When we want to study a suspension of deformable particles such as liquid droplets, macromolecules or cells, we have to account for the term $\mu(u_i n_j + u_j n_i)$ in Equation (9.15). The suspension configuration now depends also on the particle deformation which is itself a function of the local flow. The problem becomes so complicated that we can barely compute the suspension constitutive law to $O(c)$, that is, in the case where it is very dilute.

9.4   Highly Concentrated Suspension of Spheres

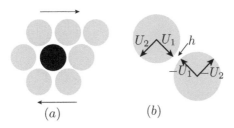

(a) (b)

FIGURE 9.3
(a) Highly concentrated suspension of solid spheres; (b) Lubrication film between two neighbouring spheres.

The configuration $\mathcal{C}$ of a very concentrated suspension is determined by packing constraints (Figure 9.3a). For example, we can assume that the centres of the spheres are arranged as a cubic system for which the maximum concentration is $c_m = 0.605$. For a concentration c slightly smaller than c_m, the thickness h of the liquid film between two spheres is very small and given by

$$h/2a = 1 - (c/c_m)^{1/3} \ll 1 \quad \text{for} \quad c \to c_m$$

Then to compute the flow around the spheres, we can use the simplifying hypotheses of lubrication theory and note that the most important contribution to energy dissipation comes from the gap between the spheres (Chapter 4). In a reference frame centred in the middle of the film, the two spheres have relative velocities $(\pm U_1, \pm U_2)$ corresponding to widening/narrowing of the film due to component U_1 or to shearing of the film due to component U_2 (Figure 9.3b).The lubrication computation (Chapter 4, Section 4.2.2) shows that the normal component U_1 is the most important one. We can thus compute the flow in the gap and obtain the apparent viscosity of a highly concentrated suspension of identical solid spheres [23]:

$$\mu_a/\mu = \frac{9}{8}\left[\frac{(c/c_{\text{max}})^{1/3}}{1-(c/c_{\text{max}})^{1/3}}\right] \tag{9.31}$$

In conclusion we note that the analytical determination of the viscosity a suspension of identical solid spheres is possible only for the two asymptotic cases corresponding to the two limits of the concentration range (Figure 9.4). There remains a domain of medium concentrations for which we must resort to numerical models.

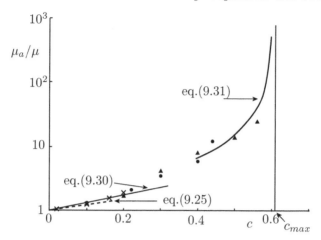

FIGURE 9.4

Apparent viscosity of a suspension identical solid spheres as a function of the volume concentration c. Symbols: experiments.

9.5   Numerical Modelling of a Suspension

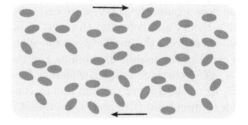

FIGURE 9.5

Modelling the motion of N particles in a volume V.

The numerical modelling of a suspension is a difficult problem. If we wanted an exact solution, we would have to follow the motion of a population of N particles in a volume V and compute at each time step the configuration and the flow field in the fluid. This would necessitate too much memory and computer time to be really efficient. Approximations must then be made, and this has led to the method called 'Stokesian dynamics' which was initially proposed by Brady and Bossis [12, 13] and which is now widely used for a number of different applications by Brady's group and others. The principle of the method consists in adding independently the contributions from the far-field flow due to each particle and the contribution from the hydrodynamic interactions between a pair of neighbouring particles. The procedure avoids the computation of the detailed flow field around each particle. This very powerful method is based on the fundamental concepts governing Stokes flows. It is thus interesting to present the principles that are used, while keeping in mind that their numerical implementation is not trivial. In the following presentation, only hydrodynamic forces are accounted for while other interaction forces (Brownian motion, electrostatic effects, etc) are neglected. We also neglect the external body forces acting on the particles, although Stokesian dynamics allow the study of such forces.

We consider a population of N particles freely suspended (with no acting external force or torque) in a Newtonian liquid, subjected to a shear flow $\mathbf{e}^\infty = <\mathbf{e}>$ (Figure 9.5). The stress tensor in this volume is thus

$$<\boldsymbol{\sigma}> = -P\mathbf{I} + 2\mu\mathbf{e}^\infty + \boldsymbol{\Sigma}^p$$

where the particle stress is due to hydrodynamic interactions effects and can be written as

$$\boldsymbol{\Sigma}^p = <\mathbf{S}>$$

where $\mathbf{S}$, called (not quite correctly) the 'stresslet', represents the integral

$$S_{ij} = \int_{S_p} \frac{1}{2}\left[\sigma_{ik}n_k x_j + \sigma_{jk}n_k x_i - \frac{2}{3}\delta_{ij}\sigma_{mk}n_k x_m\right] dS \qquad (9.32)$$

9.5.1 Global Mobility and Resistance Tensors

The starting point is to use the linearity property of the Stokes equations and to express the forces on each particle in terms of resistance tensors (Chapter 6)

$$\begin{aligned}
\mathbf{F} &= -\mathbf{R}_{FU}(\mathbf{x}) \cdot (\mathbf{U} - \mathbf{U}^\infty) + \mathbf{R}_{FE}(\mathbf{x}) \cdot \mathbf{e}^\infty \\
\mathbf{S} &= -\mathbf{R}_{SU}(\mathbf{x}) \cdot (\mathbf{U} - \mathbf{U}^\infty) + \mathbf{R}_{SE}(\mathbf{x}) \cdot \mathbf{e}^\infty
\end{aligned} \qquad (9.33)$$

where $\mathbf{x}$ is a vector describing the position and orientation of the N particles, $\mathbf{U}$ is vector describing the translation and rotation velocity of each particle, $\mathbf{U}^\infty$ is the translation and rotation velocity of the imposed flow at the centre of each particle and $\mathbf{e}^\infty$ is a vector with the six components of the shear rate tensor evaluated at the centre of each particle. Similarly, $\mathbf{F}$ is a vector containing the resultant force and torque on each particle and $\mathbf{S}$ is another vector containing the six components of the integral (9.32) for each particle. The vectors $\mathbf{x}$, $\mathbf{U}$, $\mathbf{U}^\infty$, $\mathbf{e}^\infty$, $\mathbf{F}$ and $\mathbf{S}$ have $6N$ components.

The tensors $\mathbf{R}_{FU}, \mathbf{R}_{FE}, \mathbf{R}_{SU}$ and $\mathbf{R}_{SE}$ are generalised resistance tensors ($6N \times 6N$), with indices that correspond to the corresponding coupling phenomenon. For example, $\mathbf{R}_{FE}$ gives the force on the particle due to the imposed shear rate $\mathbf{e}^\infty$.

The relation with the simple resistance tensors of Chapter 6 is obvious. However, in Chapter 6, the fluid at infinity was at rest ($\mathbf{U}^\infty = 0$ and $\mathbf{e}^\infty = 0$) and we were only seeking the force and couple on one particle. Here the external flow field ($\mathbf{U}^\infty$, $\mathbf{e}^\infty$) exerts hydrodynamic forces on each particle, which are measured in terms of resultant forces $\mathbf{F}$ and 'stresslet' $\mathbf{S}$. It is possible to rewrite Equation (9.33) in terms of a global resistance matrix $\mathcal{R}$

$$\begin{pmatrix} \mathbf{F} \\ \mathbf{S} \end{pmatrix} = -\mathcal{R} \cdot \begin{pmatrix} \mathbf{U} - \mathbf{U}^\infty \\ -\mathbf{e}^\infty \end{pmatrix} \tag{9.34}$$

where $\mathcal{R}$ is defined by

$$\mathcal{R} = \begin{pmatrix} \mathbf{R}_{FU} & \mathbf{R}_{FE} \\ \mathbf{R}_{SU} & \mathbf{R}_{SE} \end{pmatrix}$$

Inverting Equation (9.34) allows us to define a global mobility matrix $\mathcal{M}$

$$\begin{pmatrix} \mathbf{U} - \mathbf{U}^\infty \\ -\mathbf{e}^\infty \end{pmatrix} = -\mathcal{M} \cdot \begin{pmatrix} \mathbf{F} \\ \mathbf{S} \end{pmatrix} \tag{9.35}$$

9.5.2 Application to a Suspension of Spherical Particles

The method is illustrated for a suspension of N identical spheres. The flow around one sphere can be written as a sum of singularities (Chapter 8). The lower-order singularities for a freely suspended sphere are a stresslet and a potential quadrupole. The perturbation flow field around a sphere is thus

$$\mathbf{u}^p = \mathbf{u}(\text{stresslet}) + \mathbf{u}(\text{quadrupole}) + ... \tag{9.36}$$

where the detailed expressions are given in Table 8.2. The dots represent the contribution from the higher-order singularities arising because of hydrodynamic interactions. Far from the sphere, the velocity (9.36) tends asymptotically to a stresslet velocity

$$\mathbf{u}^p = \mathbf{u}(\text{stresslet}) + O\left[(a/r)^3\right], \quad r/a \to \infty$$

where r is the distance from the centre of the sphere. In other words, to evaluate the far-field velocity created by a sphere, we can replace the sphere by a stresslet (and a stokeslet if the external force is non-zero). Equation (9.35) allows us to compute the global mobility matrix $\mathcal{M}^\infty$ that corresponds to the far field.

We now have to take into account the hydrodynamic interactions between neighbouring particles. We consider only the pair interaction which is the dominant effect. This interaction is well known for two spheres in a shear flow (Figure 9.6). In particular, we know the values of the resultant force $\mathbf{F}$ and of the stresslet $\mathbf{S}$ as a function of the relative motion of the spheres.

$$\begin{pmatrix} \mathbf{F} \\ \mathbf{S} \end{pmatrix} = -\mathcal{R}_2 \cdot \begin{pmatrix} \mathbf{U} - \mathbf{U}^\infty \\ -\mathbf{e}^\infty \end{pmatrix} \tag{9.37}$$

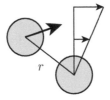

FIGURE 9.6
Hydrodynamic interaction between two spheres in a simple shear flow.

where $\mathcal{R}_2$ is the (known) tensor of resistance for a pair of spheres for any separation r/a of their centres.

We can now compute approximately the global resistance tensor for the N spheres:

$$\mathcal{R} \cong (\mathcal{M}^\infty)^{-1} + \Sigma\mathcal{R}_2 - \Sigma\mathcal{R}_2^\infty = \begin{pmatrix} \mathbf{R}_{FU} & \mathbf{R}_{FE} \\ \mathbf{R}_{SU} & \mathbf{R}_{SE} \end{pmatrix} \qquad (9.38)$$

The term $\Sigma\mathcal{R}_2$ accounts for all the pair interactions. We must thus subtract the far-field contribution of the pair interaction $\Sigma\mathcal{R}_2^\infty$ which is already accounted for in $(\mathcal{M}^\infty)^{-1}$.

The numerical procedure is then the following: starting with a given configuration $\mathcal{C}(t)$ at time time t, we replace the spheres with stresslets, compute the resulting flow field and the global mobility matrix $\mathcal{M}^\infty$ and invert the latter. We then compute the pair interaction matrix $\mathcal{R}_2$ and then the global resistance matrix from Equation (9.38). Inverting Equation (9.34) with $\mathbf{F} = 0$ allows us to find $\mathbf{U}$ and $\mathbf{S}$ for each particle:

$$\begin{aligned} \mathbf{U} &= \mathbf{U}^\infty + \mathbf{R}_{FU}^{-1} \cdot [\mathbf{R}_{FE}(\mathbf{x}) \cdot \mathbf{e}^\infty] \\ \mathbf{S} &= -\mathbf{R}_{SU}(\mathbf{x}) \cdot (\mathbf{U} - \mathbf{U}^\infty) + \mathbf{R}_{SE}(\mathbf{x}) \cdot \mathbf{e}^\infty \end{aligned} \qquad (9.39)$$

Using Equation (9.39), we can then compute the particle stress $\mathbf{\Sigma}^p = <\mathbf{S}>$ and deduce the apparent viscosity of the suspension at time t. The new configuration $\mathcal{C}(t + \Delta t)$ for the next time step is obtained by moving the particles

$$\mathbf{x}(t + \Delta t) = \mathbf{x}(t) + \mathbf{U}\Delta t$$

We then start again the computation.

The method has first allowed us to study suspensions of spheres at different concentrations with purely hydrodynamic interactions [12]. For colloidal suspensions ($a < 1\,\mu\text{m}$), we must account for Brownian motion as measured by the Péclet number (9.29). It is then found that the apparent viscosity of the suspension depends not only on the concentration, but also on the shear rate as measured by Pe (Figure 9.7). The numerical model thus allows us to understand why the suspension exhibits non-Newtonian properties as observed experimentally. Indeed, the low Pe decrease of viscosity with shear rate is due to the competition between the tendency towards particle alignment imposed by the flow and the tendency towards disorder due to Brownian motion. As the the flow force increases, the Brownian motion becomes relatively less important. The increase of viscosity with the shear rate for large Pe is due to the formation of particle clusters.

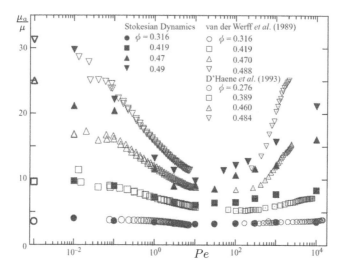

FIGURE 9.7
Apparent viscosity of a colloidal sphere suspension as a function of the Péclet number Pe, for different concentrations. Comparison between experimental and numerical results. (Reproduced from Foss and Brady [22], with permission from Cambridge University Press.)

9.6   Conclusion

Suspensions of solid or deformable particles have complex flow properties linked to the motion of their microstructure. In this chapter, we have considered mostly non-colloidal suspensions of solid particles. We have noted that even this 'simple' case was complicated! There are of course many more complex phenomena that have been measured or modelled only recently (or which are still open problems), such as

- Wall effects which lead to particle migration and thus to a non-homogeneous particle repartition

- Particle size repartition and/or particle anisotropy which influence the mechanical properties of the suspension

- Colloidal forces (Brownian motion, electrostatic interactions, steric effects due to macromolecules adsorbed on the particles, etc) which lead to a non-Newtonian behaviour of the suspension where the apparent viscosity depends on the shear rate (Chapter 11)

- Particle deformability (droplets, macromolecules, cells) which also leads to a non-Newtonian behaviour of the suspension with macroscopic anisotropy effects linked to a preferred orientation of the particles. These suspensions also exhibit a viscoelastic behaviour where the apparent viscosity depends not only on shear rate, but also on time (Chapter 11)

The study of well-characterised model suspensions allows us to identify the main physical phenomena that influence the bulk constitutive behaviour of the suspension. However, the correlation between these fundamental results and measurements on real industrial complex liquids is difficult, due to the complicated (and often secret) composition of industrial mixtures. In that case, it is often best to postulate a priori a phenomenological constitutive law, with a form guided by what is known about the fluid microstructure and by experimental observation (see Chapter 11).

9.7   Problems

9.7.1   Constitutive Law of a Suspension of Spheres

From an Ecole Polytechnique problem written with A. Poitou

The objective of the problem is the determination of a constitutive law for a suspension of identical solid spheres (radius a) freely suspended in a Newtonian liquid (viscosity μ). There are n_p particles per unit volume and the volume concentration is c. Recall that a *dilute* suspension of identical solid spheres behaves as a Newtonian fluid with apparent viscosity (Einstein's formula)

$$\mu_a = \mu \left(1 + \frac{5c}{2} \right)$$

1. Semi-dilute model

   The objective is to extend Einstein's formula to *semi-dilute* suspensions. We thus make the following assumptions

   - The bulk behaviour of the suspension is that of a Newtonian liquid with equivalent viscosity μ_a.
   - In order to account qualitatively for hydrodynamic interactions, we assume that each sphere behaves as if it were alone in a Newtonian liquid of viscosity μ_a instead of μ, the suspending fluid viscosity.

   Show that

   $$\mu_a = \frac{\mu}{1 - 2.5\,c}$$

2. Differential model

   The objective is now to extend Einstein's formula to non-dilute suspensions. We thus assume that spheres are progressively added to the suspension.

   (a) i. A volume V of suspension has concentration c_1. Find the volume of particles in V.

   ii. We now consider a suspension of particles with concentration c. We add particles to this suspension and obtain a liquid which can be considered a suspension of particles with concentration c_2 in a suspension of concentration c. Find the volume of particles in V.

   iii. Show that we can consider a particle suspension with concentration c_1, as a suspension of particles with concentration c_2 in a suspension of particles with concentration $c < c_1$ such that

   $$c_2 = \frac{c_1 - c}{1 - c}$$

(b) We start with a suspension of concentration c and equivalent viscosity $\mu_a = \mu_a(c)$. We add a small volume of particles which increases the concentration by dc. We treat the resulting suspension as a dilute suspension with concentration dc in a homogeneous liquid with viscosity $\mu_a(c)$. Compute the increase $d\mu_a$ of the apparent viscosity.

(c) Find the expression of the apparent viscosity $\mu_a(c)$. Discuss the limits of the three models (Einstein, semi-dilute and differential).

9.7.2 Intrinsic Convection in a Suspension

From an Ecole Polytechnique problem written with E. Guazzelli

This problem deals with wall effects in a settling suspension. We consider the beginning of the sedimentation process, so that the suspension can still be considered as being homogeneous. We show the apparition of a recirculation flow in the reservoir.

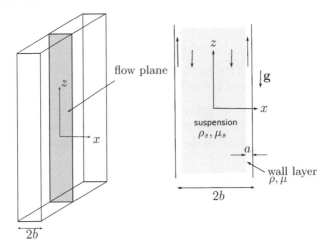

FIGURE 9.8
(a) Parallelepipedal flow cell; (b) schematics of the flow in a xz-plane.

A dilute suspension of rigid spheres (radius a, density $\rho + \Delta\rho$, $\Delta\rho > 0$) in a Newtonian incompressible liquid (viscosity μ, density ρ) is contained in a parallelepipedal cell with vertical walls and a horizontal bottom (Figure 9.8). In addition to the sedimentation motion of the particles, there is a global convection motion that we wish to study in detail. This convection originates in a particle-free layer near the walls which is due steric exclusion effects. This layer, being lighter than the suspension, has an upward motion which is compensated by the downward motion of the suspension in the centre of the cell. We position ourselves far from the bottom or the top of the cell and make the following assumptions:

- The cell thickness $2b$ is much smaller than the other dimensions (height and width). The flow is thus two-dimensional in the xz-plane (Figure 9.8), where the vertical z-axis is at the same distance from the walls

- The sphere radius is much smaller than the cell thickness $a \ll b$

- The sphere volume fraction c is small. The suspension can be considered an incompressible homogeneous liquid with density ρ_s and viscosity $\mu_s = \mu[1 + O(c)]$

- The particle-free wall layer has an average thickness a

- The motion of the suspension and of the suspending liquid is governed by the Stokes equations

$$\nabla \cdot \mathbf{u} = 0 \quad \text{and} \quad -\nabla p + \mu(\mathbf{x})\nabla^2 \mathbf{u} + \rho(\mathbf{x})\mathbf{g} = 0$$

where $\mathbf{g}$ is the gravity, and where $\rho(\mathbf{x})$ and $\mu(\mathbf{x})$ are the density and viscosity of the suspending liquid (ρ, μ) in the wall layer or of the suspension (ρ_s, μ_s) elsewhere

- The flow is quasi unidirectionnal in the z-direction $\mathbf{u} = w(x)\mathbf{e}_z$

- The velocity of the particle-free layer and of the suspension vanishes at the bottom of the cell

We denote the velocity fields in the wall layer and in the central suspension as $w^{(f)}(x)$ and $w^{(s)}(x)$, respectively. The pressure depends only on z because there is no motion normal to the vertical walls and thus no horizontal pressure gradient.

1. Compute the density ρ_s of the suspension in terms of ρ, $\Delta\rho$ and the volume concentration c of particles in the suspension.

2. Give the boundary conditions between the suspension and the wall layer at $x = \pm(b - a)$.

    (a) Find a condition that relates $w^{(f)}(x)$ and $w^{(s)}(x)$.
    (b) Find a condition for the stress. Use an order of magnitude argument to show that

    $$\frac{dw^{(f)}}{dx} \cong 0 \quad \text{at} \quad x = \pm(b - a)$$

3. We now study the flow in a particle-free layer defined by $b - a \le x \le b$. We assume that the pressure in the layer is equal to the hydrostatic pressure in the suspension $p = -\rho_s g z + \text{Cst}$.

(a) Show that the equation of motion is simply

$$\mu \frac{d^2 w^{(f)}}{dx^2} = -c \, \Delta \rho \, g$$

(b) Compute the velocity $w^{(f)}(x)$ in the layer.

(c) Show that the velocity on the edge $x = b - a$ of the particle-free layer is given by $w^* = 9cU_0/4$, where U_0 is the Stokes velocity that would have a sphere if it were alone in the fluid.

4. We now turn to the flow in the suspension. In order to simplify the computation, we neglect the thickness of the wall layer and consider that the suspension fills the cell $(-b \le x \le b)$. The suspension is driven upwards by the velocity w^* applied at $x = \pm b$ to first order. The downwards flow is driven by a pressure gradient $G = dp/dz$ which is slightly different from the hydrostatic gradient that was used for the flow computation in the wall layer.

(a) While keeping in mind that the cell has a bottom (!), find the value of the flow rate $\int_{z=\text{Cst}} w(x)dx$ through a horizontal cross section.

(b) Compute the velocity $w^{(s)}(x)$ in the suspension in terms of w^* and G.

(c) Compute G. Show that the correction to the hydrostatic gradient is $O(ca^2/b^2)$ and that we can thus neglect the viscosity difference between the pure fluid and the suspension provided the suspension is dilute $(c \ll 1)$.

(d) Compute the velocity maximum at the centre of the cell in terms of c and U_0. Draw approximately the velocity profile in a plane $z = \text{Cst}$. Do not forget to show the flow in the particle-free layer as well!

5. (a) Discuss the approximations that were made and specifically the one regarding the value of the pressure gradient in the wall layer.

(b) Under which conditions is this model valid?

(c) Which phenomena were not accounted for?

10

$O(Re)$ Correction to Some Stokes Solutions

CONTENTS

The Stokes equations have been obtained in the asymptotic case where $Re \ll 1$. However, the condition $Re \ll 1$ is quite vague and it thus desirable to have an idea of the error we make when we use the Stokes approximation, instead of the full Navier–Stokes equations, to model a flow with a given (small!) value of the Reynolds number. In order to answer this question, we consider two generic cases for which the first-order correction $O(Re)$ to the Stokes solution can be obtained in a fairly simple way. These cases deal with the uniform flow around a sphere or around an infinite cylinder.

10.1   Translation of a Sphere: Oseen Correction

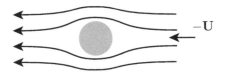

FIGURE 10.1
Sphere moving with constant velocity $\mathbf{U}$ in a fluid at rest at infinity: in a reference frame linked to the sphere, the fluid has uniform velocity $-\mathbf{U}$ at infinity.

A solid sphere with radius a moves with constant velocity $\mathbf{U}$ in a fluid at rest far from the sphere. We use a reference frame linked to the sphere and centred

on it. The fluid at infinity has thus a uniform velocity field $-\mathbf{U}$ (Figure 10.1).
In Chapter 6, Section 6.3, we found that the flow velocity $\mathbf{u}$ around the sphere
was the sum of a stokeslet $\mathbf{u}_S$ and of a potential dipole $\mathbf{u}_D$:

$$\mathbf{u} = \mathbf{u}_S + \mathbf{u}_D \qquad (10.1)$$

with orders of magnitude $O(Ua/r)$ for $\mathbf{u}_S$ and $O(Ua^3/r^3)$ for $\mathbf{u}_D$. This solution
was obtained under the condition that inertia terms were negligible compared
with the viscous terms $(Re = Ua/\nu \ll 1)$.

Far from the sphere, the velocity field is

$$\mathbf{u} = -\mathbf{U} + O(Ua/r) \quad \text{for} \quad r \gg a \qquad (10.2)$$

where the small correction to the uniform flow is due to the stokeslet. Using
Equation (10.2) for $\mathbf{u}$, we compute again the orders of magnitude of the different terms in the Navier–Stokes equations, noting that the variations of $\mathbf{u}$
are due to the stokeslet correction

$$\text{inertia} : |\mathbf{u} \cdot \nabla \mathbf{u}| \sim U \frac{Ua}{r^2} = \frac{U^2 a}{r^2}$$

$$\text{viscosity} : |\nu \nabla^2 \mathbf{u}| \sim \frac{\nu U a}{r^3}$$

The ratio between the inertia and viscosity terms is thus

$$\frac{|\mathbf{u} \cdot \nabla \mathbf{u}|}{|\nu \nabla^2 \mathbf{u}|} \sim \frac{Ua}{\nu} \left(\frac{r}{a}\right) = Re \frac{r}{a} \qquad (10.3)$$

When we move away from the sphere, r/a increases and the ratio between
inertia and viscosity becomes $O(1)$ at a distance $r = aO(Re^{-1})$ from the
sphere centre. Consequently, at this distance from the sphere (which becomes
larger as Re becomes smaller), the Stokes approximation is no longer valid.

Far from the sphere, that is, for $r \geq aO(Re^{-1})$, the fluid velocity can be
written as

$$\mathbf{u} = -\mathbf{U} + \mathbf{v} \quad \text{with} \quad |\mathbf{v}| \ll |\mathbf{U}| \qquad (10.4)$$

We replace $\mathbf{u}$ with this value in the Navier–Stokes equations (instead of the
Stokes equations) and perform the computation to first order in $|\mathbf{v}| / |\mathbf{U}|$. We
then obtain the Oseen equation

$$-(\mathbf{U} \cdot \nabla) \, \mathbf{v} = -\frac{1}{\rho} \nabla p + \nu \nabla^2 \mathbf{v} \qquad (10.5)$$

which is a first-order linear differential equation for the unknown field $\mathbf{v}$, with
a boundary condition at infinity

$$\mathbf{v} \to 0 \quad \text{for} \quad \frac{r}{a} Re \to \infty$$

and a matching condition with the Stokes solution for $a \ll r \ll a/Re$.

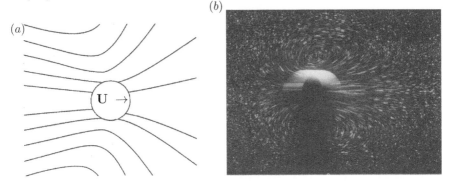

FIGURE 10.2
Flow around a sphere for $Re \cong 3.5$ in a reference frame linked to the fluid at infinity: (a) streamlines obtained from the Oseen solution; (b) experimental streamlines. (Reproduced from Coutanceau [18] with permission from Académie des Sciences de Paris). We note that the flow is no longer symmetric and that the reversibility property is thus lost.

The problem can thus be solved by means of the matched asymptotic expansion technique [42]. In particular, we find that the flow loses its upstream/downstream symmetry and thus its reversibility property, as shown in Figure 10.2. After some long computations [36], we obtain the drag force on the sphere:

$$\mathbf{F} = -6\pi\mu a\mathbf{U}\left[1+\frac{3}{8}Re+\frac{9}{40}Re^2(\ln Re)+...\right] \tag{10.6}$$

where the second term of the expansion was obtained by Oseen in 1910 and the third by Proudman and Pearson in 1957 [47]. The drag coefficient C_D is defined by

$$C_D = \frac{|\mathbf{F}|}{\frac{1}{2}\pi a^2 \rho U^2}$$

For Stokes flow, it is easy to verify that $C_D = 12/Re$ (note that the alternative form $C_D = 24/Re'$ corresponds to a Reynolds number $Re' = dU/\nu$ based on the sphere radius d). The evolution of C_D with Re is shown in Figure 10.3. Compared with experimental results, the Stokes solution is valid within 1% up to $Re \approx 0.5$. With the same precision, the Oseen solution is valid up to $Re \approx 1 - 1.5$.

The energy dissipation Φ depends on Re:

$$\Phi(Re) = -\mathbf{F} \cdot \mathbf{U} = 6\pi\mu a U^2 \left[1 + \frac{3}{8}Re + \frac{9}{40}Re^2(\ln Re) + ...\right] \tag{10.7}$$

As expected, the energy dissipation is minimum for $Re = 0$, that is, for Stokes flow, as demonstrated in Chapter 2, Section 2.5.

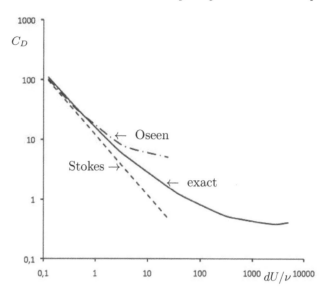

FIGURE 10.3

Drag coefficient of a sphere as a function of Reynolds number. Comparison of the Stokes and Oseen approximations with an exact solution of the Navier–Stokes equations.

10.2   Translation of a Cylinder: Stokes Paradox

A cylinder with a circular cross section of radius a moves with the constant velocity **U**, normal to its axis, in a fluid at rest. We use a reference frame linked to the cylinder and centred on its axis (Figure 10.4). The x_2-direction is the cylinder axis and **U** is along the x_3-direction. In this reference frame, the fluid at infinity has velocity $-U\mathbf{e}_3$. The Reynolds number is assumed to be small,

$$Re = Ua/\nu \ll 1$$

The flow around the cylinder is two-dimensional and can be computed with the technique developed in Chapter 3, where a stream function Ψ is introduced. It is also possible to use a singularity method with 2D singularities. Altogether, we find that the flow around the cylinder is the sum of a stokeslet and a potential dipole, given in 2D form by

$$u_i = A_i - S_i' \ln r + \frac{S_m' x_m x_i}{r^2} - \frac{T_i'}{r^2} + \frac{2T_m' x_m x_i}{r^4} \qquad (10.8)$$

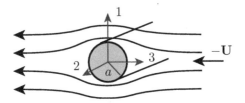

FIGURE 10.4
Cylinder moving with velocity **U** normal to its axis (direction $\mathbf{e}_2$) in a fluid at rest at infinity. We use a reference frame linked to the cylinder and centred on its axis.

It is easy to check that the above solution does indeed satisfy the 2D Stokes equations. The associated boundary conditions are

$$u_i \to -U_i \quad \text{for} \quad r \to \infty \tag{10.9}$$

and

$$u_i = 0 \quad \text{for} \quad r = a \tag{10.10}$$

The condition (10.10) is a vector equation which is evaluated on a surface. To solve Equation (10.10), we project it on the normal and tangential directions to the surface. We then obtain two equations which correspond to the two components of condition (10.10):

$$u_i n_i = 0 \quad \text{or} \quad A_i - S_i' \ln a + S_i' + \frac{T_i'}{a^2} = 0 \tag{10.11}$$

and

$$u_i - (u_k n_k) n_i = 0 \quad \text{or} \quad A_i - S_i' \ln a - \frac{T_i'}{a^2} = 0 \tag{10.12}$$

We thus have two homogeneous equations for three unknown tensors **A**, **S'** and **T'**. We can write the solution of Equations (10.11) and (10.12) in the form

$$S_i' = 2BU_i, \quad T_i'/a^2 = -BU_i, \quad A_i = 2BU_i \left(\ln a - 1/2 \right) \tag{10.13}$$

where B is a constant to be determined. The velocity field then becomes

$$u_i = BU_i \left(-2\ln \frac{r}{a} - 1 + \frac{a^2}{r^2} \right) + 2B \frac{U_m x_m x_i}{r^2} \left(1 - \frac{a^2}{r^2} \right) \tag{10.14}$$

It is clear that the velocity given by Equation (10.14) cannot satisfy condition (10.9) at infinity due to the presence of the logarithm term. It is thus impossible to find a Stokes solution to the flow around a cylinder, no matter how low the Reynolds number is! This is the *Stokes paradox*.

The flow around a cylinder does exist even when Re is small. In order to find the velocity field, we first note that for $r/a = O(Re^{-1})$, inertia and viscosity forces have the same order of magnitude. We thus introduce Re into Equation (10.14):

$$u_i = BU_i\left[-2\ln\left(Re\frac{r}{a}\right)+2\ln Re - 1 + \frac{a^2}{r^2}\right] + 2B\frac{U_m x_m x_i}{r^2}\left(1 - \frac{a^2}{r^2}\right) \quad (10.15)$$

and determine the asymptotic form of **u** for $r/a = O(Re^{-1})$:

$$u_i \sim 2BU_i \ln Re \quad \text{for} \quad r/a = O(Re^{-1}) \quad (10.16)$$

To satisfy Equation (10.9), we must have

$$B = -\frac{1}{2\ln Re} \quad (10.17)$$

which completes the determination of the velocity field.

The drag force **F** per unit axial length of cylinder is given by the stokeslet

$$\mathbf{F} = 4\pi\mu S_i' = -\frac{4\pi\mu}{\ln(1/Re)}\mathbf{U} \quad (10.18)$$

This expression was first derived by Proudman and Pearson [47].

The Stokes paradox is due to the assumption of a two-dimensional flow. Indeed, the Stokes equations are valid in the domain where $r < aO(1/Re)$. It is possible to find such a fluid domain when the object has finite dimensions (such as the sphere). For an 'infinite' cylinder, there will always be a fluid domain where $r < aO(1/Re)$ cannot be satisfied. This is why we cannot find a Stokes solution to the two-dimensional flow around a cylinder.

10.3 Validity Limits of the Stokes Approximation

The Oseen correction term to the drag force on a translating sphere gives approximately the validity limit of the Stokes approximation in this case, depending on the required precision. We note that for values of Re up to unity, the error is not very large when we use the Stokes rather than the Navier–Stokes equations. This result confirms that of Chapter 3 where we noted that the error on the streamlines in a corner was small up to values of Re of order 1 to 2.

There are many situations where the Reynolds number is small enough for the Stokes approximation to be used with a negligible error. When $Re = O(1)$, it is useful to evaluate the main physical phenomena in the system and compare them to inertia effects. We should always keep in mind that the use of the linear Stokes equations simplifies the solution of complicated problems in particular in the presence of deformable interfaces.

10.4 Problem

10.4.1 Evaluation of the Reynolds Number

The objective of the problem is to calculate under which conditions we can use the Stokes equations for the flow around different particles.

1. A solid sphere ($\rho_s = 2 \times 10^3$ kg/m^3) falls in water ($\rho = 10^3$ kg/m^3, $\mu = 10^{-3}$ Pa s) under the effect of gravity. Compute its fall velocity in terms of its radius a. Compute the value of the Reynolds number and discuss the result.

2. A solid sphere with radius $100\,\mu$m is freely suspended in water and subjected to a shear rate $\dot{\gamma}$. Compute the Reynolds number in terms of $\dot{\gamma}$ and discuss the result.

3. A gas bubble with radius 0.5 mm is suspended in water. Compute its rise velocity and the associated Reynolds number. Discuss the result.

4. A gas bubble remains spherical if the product $ReCa \ll 10^2$. Compute the capillary number Ca for a bubble with radius 0.5 mm and discuss its shape.

5. Do the same computation for a gas bubble with radius 0.5 cm in water and discuss the result.

11
<hr>

Non-Newtonian Fluids

CONTENTS

11.1   Introduction

Most natural and industrial fluids are not pure and contain particles of various nature and size. On may cite, for example, rivers, oil extraction muds, food products, cosmetic or pharmaceutical emulsions, paint, blood, polymer solutions, etc. The list is probably infinite. Such fluids which have a microstructure formed by inclusions are called *complex fluids*. Their flow behaviour does not usually satisfy Newton's constitutive law (1.12) and often exhibits one or more *non-Newtonian* effects such as

A viscosity which depends on the shear rate: Many liquids, termed 'pseudoplastic', have a viscosity which decreases as the shear rate increases. This decrease can be very significant. On the other hand, the viscosity of starch suspensions increases with the shear rate. These phenomena must obviously be accounted for in the design of transport systems for such liquids.

Unequal normal stress under simple shear flow: This effect is observed when raw egg white climbs on the whip as we rotate it, instead of moving away from the whip as water would (so-called Weissenberg effect).

Delayed response to sudden variations in flow conditions: In many liquids such as polymers, the viscous stress takes a finite time to reach a steady value after application of a constant shear rate. This phenomenon is known as *viscoelasticity*.

History effects: For given flow conditions, the stress in the fluid depends on the way the motion has been set.

The composition and the microstucture of those fluids control the flow properties. The determination of the flow properties is a crucial issue for the manipulation or the quality control of non-Newtonian fluids. However, the complexity of the microstructure usually prevents us from using the fundamental approach developed for suspensions. The first attempts to model the behaviour of non-Newtonian fluids were thus purely empirical and consisted of

- The addition of non-linear terms to Newton's law (1.12) in order to describe the variation of viscosity with shear rate

- The addition of an elastic term to Equation (1.12), to account for viscoelasticity, etc

Since the years 1950–1960 (theoretical work of Oldroyd, of Truesdell and of Rivlin and Noll), a fundamental theory has been established for the proper formulation of consistent constitutive laws allowing us to model a large range of different behaviours.

Schematically, we can divide the research on non-Newtonian fluids into two main domains:

- *Rheology* aims to find an appropriate constitutive law and to determine its parameters for describing the behaviour of a fluid or of a class of fluids under a range of flow conditions. This has led to the design of different measuring devices. In all cases, the objective is to create a shear flow which is simple enough to be analysed and to allow measuring the material parameters of the fluid. Most of these devices use a small sample of fluid, which thus limits the value of the Reynolds number. It is then usual to neglect the inertia terms, which simplifies the analysis of the flow field. The most common flow field is the linear simple shear flow. Rheology is particularly important for the quality control of fluids since the bulk properties reflect the state of the microstructure.

- *Non-Newtonian fluid mechanics* cover the solution of various flow problems involving non-Newtonian liquids. One may cite, for example, mould filling, computation of paint coating films or polymer extrusion. We obviously first need a general constitutive law with coefficients that can be determined from simple experiments. The field of application of complicated laws has been often limited to simple flow situations, but has been recently extended with the extensive use of numerical modelling.

11.2 Non-Newtonian Fluid Mechanics

11.2.1 Equations of Motion

In the following, the fluids are always incompressible:

$$\text{tr } \mathbf{e} = 0 \quad \text{or} \quad \nabla \cdot \mathbf{u} = 0 \tag{11.1}$$

We introduce the stress deviator $\boldsymbol{\tau}$ defined by

$$\boldsymbol{\tau} = \boldsymbol{\sigma} - 1/3\text{tr}\,(\boldsymbol{\sigma})\mathbf{I} \tag{11.2}$$

which is equivalent to

$$\boldsymbol{\sigma} = -p\mathbf{I} + \boldsymbol{\tau} \tag{11.3}$$

where $\text{tr}(\boldsymbol{\tau}) = 0$ by definition. The stress deviator for a Newtonian fluid is obviously

$$\boldsymbol{\tau} = 2\mu\mathbf{e} \tag{11.4}$$

For a non-Newtonian fluid, the momentum equation is

$$\rho\frac{D\mathbf{u}}{Dt} = \rho\mathbf{g} - \nabla p' + \nabla \cdot \boldsymbol{\tau} \quad \text{or} \quad \rho\frac{Du_i}{Dt} = \rho g_i - \frac{\partial p'}{\partial x_i} + \frac{\partial \tau_{ij}}{\partial x_j} \quad (11.5)$$

If gravity is included into the pressure term, the equation of motion becomes

$$\rho\frac{D\mathbf{u}}{Dt} = -\nabla p + \nabla \cdot \boldsymbol{\tau} \quad \text{or} \quad \rho\frac{Du_i}{Dt} = -\frac{\partial p}{\partial x_i} + \frac{\partial \tau_{ij}}{\partial x_j} \quad (11.6)$$

To these equations of motion (11.1) and (11.6), we must add the constitutive law of the fluid and the boundary conditions.

11.2.2 Formulation of a Constitutive Law

A constitutive law relates the stress in the fluid to the motion (deformation, rate of deformation, etc). We consider only mechanical interactions. The principles which a constitutive law must satisfy are stated in the following:

Determinism: The stress in a material depends only on the *past* history of deformation to which it has been subjected (in other words, a material cannot predict the future).

Second principle of thermodynamics: The constitutive law must satisfy the second principle of thermodynamics. This condition puts restrictions on the sign of some of the coefficients of a law (e.g. the viscosity of a Newtonian fluid is non-negative).

Local action principle: The stress at a point $\mathbf{x}$ depends only on the interaction with neighbouring particles. This hypothesis excludes long-range interactions between particles (e.g. electrostatic forces). When they are present, such forces can eventually be accounted for as volume forces.

Invariance principle: The constitutive law must be invariant under a change of coordinate system. This is ensured by writing the law in tensor form.

Objectivity principle: The material properties of a fluid are intrinsic properties. Thus, the resulting effect of some action must be independent of the observer: the constitutive law must be invariant under a change of reference frame and of origin of time.

There exists in the literature a large number of constitutive laws based on some physical assumptions. The objective is to have a law which describes well the experimentally observed behaviour and which is mathematically simple enough to allow computing the flow. Furthermore, it helps if the material constants that appear in the law can be determined experimentally!

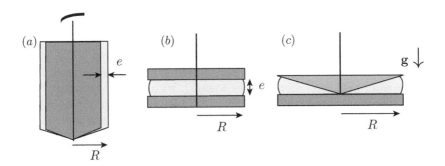

FIGURE 11.1
Different shear viscometers: (*a*) Couette viscometer with vertical co-axial
cylinders; (*b*) plane–plane viscometer, consisting of two horizontal parallel
co-axial disks; (*c*) cone–plane viscometer, consisting of a cone co-axial with a
disk. The flow in the devices is a simple shear flow when the gap *e* is small
compared with the radius *R*. The shear rate $\dot{\gamma}$ is uniform is cases (*a*) and (*c*),
and varies with the distance to the rotation axis in case (*b*).

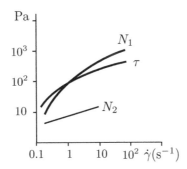

FIGURE 11.2
Typical viscometric functions as a function of shear rate for a polymer solution.

11.2.3 Viscometric Parameters in Simple Shear Flow

In order to characterise the behaviour of the fluid, we subject it to a simple shear flow defined in Cartesian coordinates by

$$u_1 = \dot{\gamma} x_2, \quad u_2 = u_3 = 0 \tag{11.7}$$

where $\dot{\gamma}$ is the shear rate. This type of flow can be created in a viscometer with co-axial cylinders (Couette viscometer), with parallel co-axial disks or with a cone co-axial to a disk (Figure 11.1). In all cases, the device axis is vertical. The flow in such viscometers is nearly a simple shear flow inasmuch as the gap e is small compared with the radius R; thus, if $e \ll R$ for cases (a) and (b) or if the cone angle is small (typically of order 2° or 3°) in case (c). Then in an appropriate reference system, the rate of strain tensor has only two non-zero components:

$$e_{12} = e_{21} = \dot{\gamma}/2 \tag{11.8}$$

The invariants of a tensor are quantities which have the same value in any coordinate system. Since $\mathbf{e}$ is a second-order tensor, it has three independent invariants which can be defined as

$$I_1 = \text{tr}(\mathbf{e}), \qquad I_2 = \text{tr}(\mathbf{e}^2), \qquad I_3 = \text{tr}(\mathbf{e}^3) \tag{11.9}$$

Of course, any combination of invariants is another invariant. For an incompressible liquid, the invariant $I_1 = \text{tr}(\mathbf{e})$ is always zero. For the simple shear flow (11.7), the invariants in Equation (11.9) are given by

$$I_1 = 0, \qquad I_2 = \dot{\gamma}^2/2, \qquad I_3 = 0 \tag{11.10}$$

It is possible to characterise the fluid behaviour by means of the following viscometric functions:

- The apparent viscosity defined by analogy with a Newtonian liquid as

$$\mu_a = \tau_{12}/2e_{12} = \tau/\dot{\gamma} \tag{11.11}$$

where τ is the shear stress

- The two normal stress differences

$$N_1 = \sigma_{11} - \sigma_{22} \quad \text{and} \quad N_2 = \sigma_{22} - \sigma_{33} \tag{11.12}$$

We note that for a Newtonian liquid, the apparent viscosity μ_a is constant and equal to the fluid viscosity μ, and that the normal stress differences are identically zero, $N_1 = N_2 = 0$.

Experimentally, τ, μ_a, N_1 and N_2 are fairly easy to measure. For a polymer solution, we obtain the typical results shown in Figure 11.2. A *shear thinning* behaviour is observed where the shear stress τ increases non-linearly with

$\dot{\gamma}$ and where the apparent viscosity decreases with $\dot{\gamma}$. Furthermore, the first normal stress difference N_1 is positive, whereas the second N_2 is always much smaller in absolute value than N_1, and sometimes even negative [52]. For some suspensions or for pasty liquids, we observe a transient behaviour similar to that shown in Figure 11.3 when we suddenly change the flow conditions.

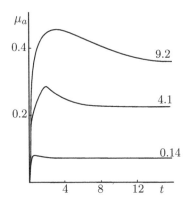

FIGURE 11.3
Transient apparent viscosity μ_a (mPa s) as a function of time t (s) in response to a sudden start of flow for a suspension of red blood cells in a fibrinogen solution with different concentrations ($9.2, 4.1, 0.14$ g/L). High fibrinogen concentrations induce the formation of cell clusters which are then dissociated by shear stress. The cluster destruction takes some time to occur. The associated energy cost leads to the apparent viscosity overshoot. (Adapted from [14]).

Note that apart from the rotating ones, there exist other types of viscometers (elongational, capillary viscometers, etc) where the flow field is different from Equation (11.7).

In conclusion, non-Newtonian liquids exhibit complicated flow properties which we now want to model with constitutive laws, as simply as possible.

11.3   Viscous Non-Newtonian Liquid

11.3.1   Reiner–Rivlin Fluid

We assume that the stress in the fluid is entirely determined by the local rate of strain at any time. The fluid response is thus instantaneous and local:

$$\boldsymbol{\tau}(\mathbf{x}, t) = \boldsymbol{\mathcal{G}}[\mathbf{e}(\mathbf{x}, t)] \tag{11.13}$$

where $\mathcal{G}$ is a tensor function. After expanding Equation (11.13) as a Taylor series and noting that $[\mathbf{I}, \mathbf{e}, \mathbf{e}^2]$ forms a basis in which we can project $\mathbf{e}^n$ for any $n \geq 3$, we obtain the simple form

$$\mathcal{G}(\mathbf{e}) = a_0\mathbf{I} + a_1\mathbf{e} + a_2\mathbf{e}^2 \tag{11.14}$$

which satisfies the objectivity principle since $\mathbf{e}$ is objective. The scalar coefficients a_i depend only on the two invariants I_2 and I_3 of $\mathbf{e}$. The stress tensor thus becomes

$$\boldsymbol{\sigma} = -p\mathbf{I} + a_1(I_2, I_3)\mathbf{e} + a_2(I_2, I_3)\mathbf{e}^2 \tag{11.15}$$

where the term a_0 has been included in the pressure p. This general equation is known as the Reiner–Rivlin law. The coefficients a_i are fluid material functions that must be determined experimentally. As a particular case, we recover the Newtonian liquid with constant viscosity μ for

$$a_2 = 0 \quad \text{and} \quad a_1 = 2\mu$$

When a Reiner–Rivlin fluid is undergoing a linear shear flow (11.7), the stress components are obtained from Equations (11.8) and (11.15):

$$\tau_{12} = \tau_{21} = a_1\dot{\gamma}/2$$

$$N_1 = \sigma_{11} - \sigma_{22} = 0 \quad \text{and} \quad N_2 = \sigma_{22} - \sigma_{33} = a_2\dot{\gamma}^2/4 \tag{11.16}$$

The Reiner–Rivlin law thus predicts that the apparent viscosity is equal to $a_1/2$ and can depend on the shear rate since a_1 is a function of the invariants of $\mathbf{e}$ and thus of $\dot{\gamma}^2$. In particular, we can model a power law behaviour

$$\tau_{12} = \tau = k\dot{\gamma}|\dot{\gamma}|^{n-1} \tag{11.17}$$

where n is the flow index and k the consistency (note that the units of k depend on the law!). Typical graphs of τ versus $\dot{\gamma}$, also called rheograms, are shown in Figure 11.4.

The corresponding apparent viscosity is

$$\mu_a = k|\dot{\gamma}|^{n-1} \tag{11.18}$$

For $n = 1$, we retrieve the Newtonian behaviour. For $n > 1$, the apparent viscosity increases with shear rate and the fluid is said to be shear-thickening or dilatant (this is the case of starch pastes, cement, tooth paste, etc). For $n < 1$, the apparent viscosity decreases when $\dot{\gamma}$ increases and the fluid is said to be shear-thinning or pseudoplastic (this is the case of polymer solutions, emulsions, etc). The power law is also called the Ostwald law when $n < 1$.

However, some problems arise for pseudoplastic fluids. Firstly, Equation (11.18) predicts that the apparent viscosity becomes infinite when the shear rate tends to zero. Secondly, the law predicts indeed that there are normal stress differences (11.16), which is a characteristic property of non-Newtonian

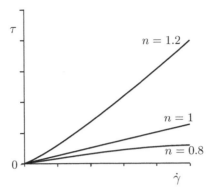

FIGURE 11.4
Rheogram of a fluid obeying a power law.

fluids. Unfortunately, no experiment up to now has ever found equal values for σ_{11} and σ_{22} when σ_{22} and σ_{33} were different. This casts doubt on the physical relevance of the Reiner–Rivlin law. However, since the power law is very simple with material coefficients k and n that are easy to determine, it is often used to analyse rheological measurements. In particular when quality control is the objective of such measurements, the power law provides an easy way to quantify rheograms.

When normal stress effects are negligible (or neglected), a simplified general form of the Reiner–Rivlin law can be used to describe the behaviour of the fluid

$$\boldsymbol{\sigma} = -p\mathbf{I} + a_1(\text{tr } \mathbf{e}^2)\mathbf{e}$$

where viscosity is a function of shear rate and more precisely of the energy dissipation in the flow.

11.3.2 Fluid with a Yield Stress

Many fluids having a cohesive microstructure need a finite stress level to be put into motion. They are called viscoplastic fluids or yield stress fluids. The physical interpretation is that, in order to set the fluid into motion, one must apply a minimum stress to dissociate the cohesive microstructures which are formed at rest. For example, blood, clay suspensions, paper paste, tar, some dressings (ketchup, mayonnaise) all have a yield stress.

The yield stress condition can be written as a von Mises criterion:

$$\mathbf{e} = 0 \qquad\qquad \text{if} \quad \tfrac{1}{2}\text{tr}(\boldsymbol{\tau}^2) \leq \tau_s^2$$

$$\boldsymbol{\tau} = \left(2\mu_p + \frac{\tau_s}{\sqrt{I_2/2}}\right)\mathbf{e} \quad \text{if} \quad \tfrac{1}{2}\text{tr}(\boldsymbol{\tau}^2) > \tau_s^2 \tag{11.19}$$

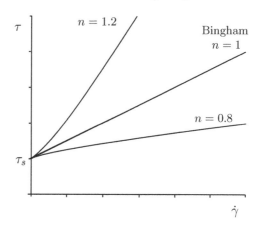

FIGURE 11.5
Rheograms of fluids with yield stress τ_s.

where τ_s is the yield stress and μ_p the plastic viscosity. The constitutive law
(11.19) is called the Bingham law. In the case of the simple shear flow (11.7),
we get simply

$$\dot\gamma = 0 \qquad \text{if} \qquad |\tau_{12}| \leq \tau_s$$

$$\tau_{12} = \mu_p \dot\gamma + \tau_s \qquad \text{if} \qquad |\tau_{12}| > \tau_s \tag{11.20}$$

The corresponding rheogram is shown in Figure 11.5 where we note that the
flow behaves linearly when the yield stress is exceeded. Bingham's law is so
simple that it is used whenever possible and appropriate. When the fluid
behaves in a non-linear way after the yield stress, we can use the Herschell–
Bulkley law which, for simple shear flow, is an empirical combination of a
power law and a yield stress (Figure 11.5)

$$\dot\gamma = 0 \qquad \text{if} \qquad \tau_{12} \leq \tau_s$$

$$\tau_{12} = k\dot\gamma^n + \tau_s \qquad \text{if} \qquad \tau_{12} > \tau_s \tag{11.21}$$

This law is often used to quantify the rheograms of pastes. The general form
is the same as Equation (11.19) where μ_p depends on $\text{tr}(\mathbf{e}^2)$.

Remarks

- Yield stress laws must be used with care when the stress is not uniform in the flow. In this case, the fluid domains that are subjected
  to a stress level below the yield stress are not sheared (but can have
  a solid-like motion such as plug flow). This situation arises in tube
  flow where the stress is zero on the tube axis and maximum at the

wall. The relation between the pressure drop and the flow rate then depends on both μ_p and τ_s.

- The value of the yield stress is relative. Indeed, the yield stress is usually found by extrapolating a rheogram to zero shear rate. The minimum value of the shear rate (the 'zero shear') depends on the precision of the viscometer. Thus we can say that the value of the yield stress also depends on the viscometer. This is not a problem if τ_s is used only to quantify experimental rheograms always obtained with the same viscometer.

11.4 Viscoelastic fluid

The behaviour of a fluid at time t often depends on the shear rate applied at earlier instants $t' < t$. This means that we must introduce a further degree of complexity, as compared to purely viscous liquids with or without a yield stress, since we have to take into account the shear rate history and the intrinsic response time of the fluid. This response time is directly linked to the physical properties of the microstructure and to its motion and deformation under the imposed flow. Such liquids are called 'viscoelastic'. Examples include polymer solutions, shampoo and emulsions.

By definition, a fluid has no preferred configuration and this is why the stress depends only on the rate of deformation rather than on the deformation itself, as is the case for a solid. The introduction of elastic effects in a fluid then raises the question of the definition of a reference state with respect to which we can measure a deformation. Truesdell has proposed a model of a viscoelastic fluid where the stress depends both on the rate of deformation (visco) and a deformation (elastic). The latter is defined as the deformation between the fluid configuration at the time t of measurement and the configuration at an earlier time t' $(t' < t)$.

11.4.1 Relative Deformation

We now show more precisely how the deformation of the fluid is measured.

Let $\mathbf{x}(t')$ be the position of a material point of the fluid at time t'. This point is labelled by its position $\mathbf{x}_t = \mathbf{x}(t)$ at the observation time $t > t'$. The equation of motion of each point is then

$$\mathbf{x}(t') = \boldsymbol{\zeta}_t(\mathbf{x}_t, t') \tag{11.22}$$

The index t indicates that we have chosen the configuration at time t as the reference state. The velocity of the point at time t' is then

$$\mathbf{u}(t') = \frac{\partial \boldsymbol{\zeta}_t(\mathbf{x}_t, t')}{\partial t'}$$

The displacement gradient $\mathbf{F}_t$ between two material points is given by

$$d\mathbf{x}(t') = \mathbf{F}_t(t') \cdot d\mathbf{x}_t \qquad (11.23)$$

This definition has two consequences:

$$\mathbf{F}_t(t) = \mathbf{I} \quad \text{and} \quad \mathbf{F}_t(t') = \nabla \otimes \boldsymbol{\zeta}_t$$

Obviously, the expression of $\mathbf{F}_t$ depends on t, which means that $\mathbf{F}_t$ is a *relative* displacement gradient. We now define the relative Cauchy dilatation tensor $\mathbf{C}_t$ which is symmetric:

$$\mathbf{C}_t = {}^{\mathsf{T}}\mathbf{F}_t \cdot \mathbf{F}_t \qquad (11.24)$$

This leads to

$${}^{\mathsf{T}}d\mathbf{x}(t') \cdot d\mathbf{x}(t') = {}^{\mathsf{T}}d\mathbf{x}_t \cdot \mathbf{C}_t \cdot d\mathbf{x}_t \qquad (11.25)$$

Similarly, we can also introduce the relative Lagrange deformation tensor ε_t:

$$\varepsilon_t = \frac{1}{2}(\mathbf{C}_t - \mathbf{I}) \qquad (11.26)$$

All these tensors are equivalent to those commonly used in continuum mechanics. The main difference is that they are measured with respect to the configuration of the medium at time t rather than with respect to a fixed reference configuration.

When we derive Equation (11.23) with respect to t', we obtain the relative velocity gradient

$$d\mathbf{u}(t') = \dot{\mathbf{F}}_t \cdot d\mathbf{x}_t = \nabla \mathbf{u} \cdot d\mathbf{x}_t \quad \text{thus} \quad \nabla \mathbf{u} = \dot{\mathbf{F}}_t \qquad (11.27)$$

To compute the rate of deformation, we derive Equation (11.24) with respect to t'

$$\dot{\mathbf{C}}_t = {}^{\mathsf{T}}\dot{\mathbf{F}}_t \cdot \mathbf{F}_t + {}^{\mathsf{T}}\mathbf{F}_t \cdot \dot{\mathbf{F}}_t.$$

For $t = t'$, we obtain

$$\dot{\mathbf{C}}_{t=t'} = \dot{\mathbf{C}}(t) = {}^{\mathsf{T}}\nabla \mathbf{u} + \nabla \mathbf{u} = 2\mathbf{e} \qquad (11.28)$$

We recover the fact that the derivative of the dilatation $\mathbf{C}$ at time t (or at any time t' if the deformation is small) is equal to twice the deformation rate $\mathbf{e}$.

11.4.2   General Constitutive Law

Consider two reference systems (1) and (2) in which a tensor has components $\mathbf{A}^{(1)}$ and $\mathbf{A}^{(2)}$. This tensor is objective if it satisfies the following transformation relation:

$$\mathbf{A}^{(1)} = \mathbf{Q} \cdot \mathbf{A}^{(2)} \cdot {}^{\top}\mathbf{Q} \tag{11.29}$$

where $\mathbf{Q}$ is the orthogonal transformation matrix that relates the coordinates in reference systems (1) and (2). For relative tensors, the transformation from system (1) to system (2) involves both $\mathbf{Q}(t)$ and $\mathbf{Q}(t')$. A relative tensor is objective if the transformation (11.29) involves $\mathbf{Q}(t)$ only, that is, the relative orientation of the two frames at time t. It is easy to verify from definitions (11.23) and (11.24) that $\mathbf{F}_t$ is not objective, whereas $\mathbf{C}_t$ is.

The principle of local action implies that $\boldsymbol{\tau}$ depends only on $\boldsymbol{\zeta}_t$ and its spatial gradients. Then the objectivity principle leads to the following general law:

$$\boldsymbol{\tau}(\mathbf{x}, t) = \overset{t}{\underset{-\infty}{\mathcal{G}}}\ [\mathbf{C}_t(\mathbf{x}, t')] \tag{11.30}$$

where $\mathcal{G}$ is a tensor functional. A fluid which obeys this law is called a 'simple fluid'. A constitutive law in the form (11.30) is much too general to be conveniently used to compute the flow of a non-Newtonian fluid. Furthermore, the present experimental devices do not allow the determination of the functional $\mathcal{G}$ for all possible histories of deformation. It is thus necessary to postulate specific forms of $\mathcal{G}$, based in particular on the physics of the microstructure. For example, a suspension of macromolecules will have elastic effects due to the elongation of the molecular chains under shear which can then be modelled by means of the kinetic theory of elastomer elasticity.

11.5   Linear Viscoelastic Laws

The *linear* viscoelastic laws are limited to *small* deformation of the material.

11.5.1   Maxwell Fluid

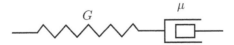

FIGURE 11.6
Maxwell model where an elastic component (spring with modulus G) and a viscous component (dashpot with viscosity μ) are added in series.

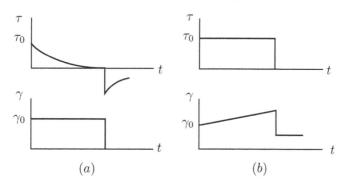

FIGURE 11.7

Transient response of a Maxwell fluid: (a) stress relaxation under constant deformation γ_0; (b) creep under constant stress τ_0.

A simple way to model the viscoelastic behaviour consists of adding the viscous and elastic responses. In the case of small deformation, the stresses $\boldsymbol{\tau}_v$ and $\boldsymbol{\tau}_e$ due to a viscous or elastic deformation, respectively, are given by

$$\boldsymbol{\tau}_v = 2\mu\,\mathbf{e}_v \quad \text{and} \quad \boldsymbol{\tau}_e = G(\mathbf{C}_t - \mathbf{I}) = 2G\,\varepsilon_t \tag{11.31}$$

where μ and G are material coefficients (with dimensions of viscosity and of elastic modulus, respectively), $\mathbf{e}_v$ is the rate of deformation tensor and ε_t is the relative deformation tensor. The elastic and viscous responses can then be associated in different fashions. For example, it is possible to assume that the deformation of a viscoelastic element is the sum of the viscous and elastic deformations under the same stress. This is shown schematically in Figure 11.6 where the Maxwell model of a viscoelastic fluid is represented by the series association of a viscous element (dashpot with viscosity μ) and of an elastic element (spring with modulus G). In the limit of small deformation, the elastic rate of strain is simply $\dot{\varepsilon}_t$. Consequently, the Maxwell model corresponds to

$$\boldsymbol{\tau}_v = \boldsymbol{\tau}_e = \boldsymbol{\tau} \quad \text{and} \quad \mathbf{e} = \mathbf{e}_v + \dot{\varepsilon}_t \tag{11.32}$$

Combining Equations (11.31) and (11.32), we obtain the Maxwell constitutive law

$$\boldsymbol{\tau} + \lambda\dot{\boldsymbol{\tau}} = 2\mu\mathbf{e} \tag{11.33}$$

where $\lambda = \mu/G$ is the characteristic response time of the fluid.

A classical experiment consists of subjecting the fluid of a simple shear flow (11.7). The only non-zero components of the deformation tensor are

$$\varepsilon_{t12} = \varepsilon_{t21} = \gamma(t)/2 \tag{11.34}$$

and the only non-zero components of the stress deviator are

$$\tau_{12} = \tau_{21} = \tau(t) \tag{11.35}$$

A relaxation measurement consists of imposing a constant deformation $\gamma(t) = \gamma_0$ at time $t = 0$, which results in an instantaneous elastic stress response $\tau_0 = G\gamma_0$ at time $t = 0$. Since ε_t is constant, $e = 0$ and the stress decreases exponentially from the initial value τ_0 (Figure 11.7a)

$$\tau(t) = \tau_0 e^{-t/\lambda} = G\gamma_0 e^{-t/\lambda}$$

When we return the deformation to zero, the stress decreases instantaneously by $-G\gamma_0$ and then goes back to zero exponentially (Figure 11.7a). A relaxation experiment is characterised by a relaxation function $\phi(t)$:

$$\phi(t) = \tau(t)/\gamma_0 = G e^{-t/\lambda} = \frac{\mu}{\lambda} e^{-t/\lambda} \qquad (11.36)$$

Conversely, we can perform a creep experiment where we apply a constant stress $\tau(t) = \tau_0$ at time $t = 0$. After an instantaneous elastic response, the fluid deformation increases linearly with time:

$$\gamma(t) = \frac{\tau_0}{G}(1 + t/\lambda)$$

When we remove the stress ($\tau = 0$), the deformation decreases instantaneously by the elastic contribution and remains constant (Figure 11.7b). The creep function $J(t)$ is defined by

$$J(t) = \gamma(t)/\tau_0 = \frac{1}{G}(1 + t/\lambda) \qquad (11.37)$$

The relaxation and creep experiments are easy to perform in a rotating viscometer where we impose either the deformation (relaxation) or the torque (creep) and measure the corresponding fluid response. The analysis of the relaxation or creep curves then allows us to determine the parameters μ and λ.

The main advantage of the Maxwell model is the simplicity with which it accounts for the time response of the fluid. However, it is limited to small deformation. Furthermore, it is easy to verify that under a constant shear rate, the fluid viscosity is constant and there are no normal stress effects.

11.5.2 Generalised Maxwell Fluid

Another limitation of the Maxwell model is that it has only one response time, which does not always allow proper matching with experimental data. A generalised Maxwell model with N different relaxation times can be obtained from the parallel association of N Maxwell models each having a specific response time λ_n. The stress is then given by

$$\boldsymbol{\tau} = \sum_{1}^{N} \boldsymbol{\tau}_n \quad \text{with} \quad \boldsymbol{\tau}_n + \lambda_n \dot{\boldsymbol{\tau}}_n = 2\mu_n \, \mathbf{e} \qquad (11.38)$$

The relaxation function is the sum of the N relaxation functions of each element:

$$\phi(t) = \sum_1^N \frac{\mu_n}{\lambda_n} e^{-t/\lambda_n}$$

It is possible to show that Equation (11.38) is equivalent to

$$\boldsymbol{\tau} + \sum_1^N a_n \frac{\partial^n \boldsymbol{\tau}}{\partial t^n} = b_0 \mathbf{e} + \sum_1^{N-1} b_n \frac{\partial^n \mathbf{e}}{\partial t^n} \tag{11.39}$$

where a_n and b_n are combinations of the coefficients λ_n and μ_n. In the simple case where there are only two relaxation times, Equation (11.39) becomes

$$\boldsymbol{\tau} + a_1 \frac{\partial \boldsymbol{\tau}}{\partial t} + a_2 \frac{\partial^2 \boldsymbol{\tau}}{\partial t^2} = b_0 \mathbf{e} + b_1 \frac{\partial \mathbf{e}}{\partial t} \tag{11.40}$$

which simplifies further if $a_2 = 0$,

$$\boldsymbol{\tau} + a_1 \frac{\partial \boldsymbol{\tau}}{\partial t} = b_0 \mathbf{e} + b_1 \frac{\partial \mathbf{e}}{\partial t} \tag{11.41}$$

This is the Jeffrey's law which corresponds to the parallel association of a Maxwell model with a dashpot. This law has two relaxation times a_1 and b_1/b_0 which correspond to the responses to applied deformation or to applied stress, respectively. It is well adapted to model pasty fluids, such as ice cream or plaster which have two response times linked to the destruction of the microstucture for one and to the deformation of the microstructure under shear for the other.

11.5.2.1 Proof of Equation (11.39)

We prove Equation (11.39) for the simple case $N = 2$. Two Maxwell models are associated in parallel:

$$\boldsymbol{\tau} = \boldsymbol{\tau}_1 + \boldsymbol{\tau}_2$$

with

$$\boldsymbol{\tau}_1 + \lambda_1 \dot{\boldsymbol{\tau}}_1 = 2\mu_1 \mathbf{e} \quad \text{and} \quad \boldsymbol{\tau}_2 + \lambda_2 \dot{\boldsymbol{\tau}}_2 = 2\mu_2 \mathbf{e}$$

We compute

$$
\begin{aligned}
(\lambda_1 + \lambda_2)\frac{\partial \boldsymbol{\tau}}{\partial t} &= \lambda_1 \frac{\partial \boldsymbol{\tau}_1}{\partial t} + \lambda_2 \frac{\partial \boldsymbol{\tau}_2}{\partial t} + \lambda_1 \frac{\partial \boldsymbol{\tau}_2}{\partial t} + \lambda_2 \frac{\partial \boldsymbol{\tau}_1}{\partial t} \tag{11.42}\\
&= 2(\mu_1 + \mu_2)\mathbf{e} - (\boldsymbol{\tau}_1 + \boldsymbol{\tau}_2) + \lambda_2 \frac{\partial \boldsymbol{\tau}_1}{\partial t} + \lambda_1 \frac{\partial \boldsymbol{\tau}_2}{\partial t}\\
&= 2(\mu_1 + \mu_2)\mathbf{e} - \boldsymbol{\tau} + \lambda_2 \frac{\partial \boldsymbol{\tau}_1}{\partial t} + \lambda_1 \frac{\partial \boldsymbol{\tau}_2}{\partial t}
\end{aligned}
$$

We also compute

$$\lambda_1\lambda_2\frac{\partial^2\boldsymbol{\tau}}{\partial t^2} = \lambda_1\lambda_2\frac{\partial^2\boldsymbol{\tau}_1}{\partial t^2} + \lambda_1\lambda_2\frac{\partial^2\boldsymbol{\tau}_2}{\partial t^2}$$

$$= 2(\lambda_2\mu_1 + \lambda_1\mu_2)\frac{\partial\mathbf{e}}{\partial t} - \lambda_2\frac{\partial\boldsymbol{\tau}_1}{\partial t} - \lambda_1\frac{\partial\boldsymbol{\tau}_2}{\partial t} \qquad (11.43)$$

Combining Equations (11.42) and (11.43), we obtain Equation (11.39) with

$$a_1 = \lambda_1 + \lambda_2, \quad a_2 = \lambda_1\lambda_2, \quad b_0 = 2(\mu_1 + \mu_2), \quad b_1 = 2(\lambda_1\mu_2 + \lambda_2\mu_1)$$

11.5.3   Linear Viscoelastic Law: Integral Form

If we now assume that the stress $\boldsymbol{\tau}(t)$ in the fluid at time t results from the additive effect of all the deformations applied at time $t' < t$, then Maxwell's law (11.33) can be written in the integral form

$$\boldsymbol{\tau}(t) = 2\int_{-\infty}^{t}\phi(t - t')\mathbf{e}(t')\,dt' \qquad (11.44)$$

where $\phi(t)$ is the relaxation function of Maxwell's law. Similarly, for the generalised model, the integral form is

$$\boldsymbol{\tau}(t) = 2\int_{-\infty}^{t}\sum_{n=1}^{N}\frac{\mu_n}{\lambda_n}e^{-(t-t')/\lambda_n}\mathbf{e}(t')\,dt' \qquad (11.45)$$

Instead of a discrete spectrum of relaxation times, it is possible to introduce a continuous one for which the relaxation function is

$$\phi(t) = \int_0^\infty \frac{H(\lambda)}{\lambda}e^{-t/\lambda}d\lambda \qquad (11.46)$$

where $H(\lambda)/\lambda$ is the relaxation time spectrum.

Finally, it is also possible to write the constitutive law in terms of the relative deformation ε_t instead of the shear rate $\mathbf{e}$. The operation is possible for small deformations only, such that $\mathbf{e}(t') = \partial\varepsilon_t(t')/\partial t'$. Integrating Equation (11.44) by parts, we find

$$\boldsymbol{\tau}(t) = 2\int_{-\infty}^{t}f(t - t')\varepsilon_t(t')\,dt' = 2\int_0^\infty f(s)\varepsilon_t(t - s)\,ds \qquad (11.47)$$

where $s = t - t'$ and where

$$f(s) = d\phi(s)/ds$$

is the memory function of the fluid. The fact that $f(s) \to 0$ for $s \to \infty$ means that the fluid has a fading memory, that is, it forgets the deformations which were applied in the distant past.

11.6   Non-Linear Viscoelastic Laws

We now relax the small deformation assumption and seek constitutive laws
which are valid under arbitrary deformations. The result is a non-linear law to
be solved numerically. However, there exist commercial computer codes that
allow solving non-Newtonian fluid mechanics problems with different constitutive laws.

11.6.1   Non-Linear Integral Laws

In a first approach, we decouple the temporal and mechanical effects and assume that the stress in the fluid is the result of all past deformations, weighted
by a memory function $f(s)$ where $s = t - t'$ is the memory variable. We thus
obtain for example the Green–Rivlin law

$$\boldsymbol{\tau}(t) = \int_0^\infty f(s)\left[\mathbf{C}_t(t-s) - \mathbf{I}\right] ds \tag{11.48}$$

For small deformations Equation (11.48) is indeed the limit of Equation
(11.47). However, for finite deformations, there is no unique way to define
the 'deformation'. For example, we can use either $\mathbf{C}_t$ or $\mathbf{C}_t^{-1}$. But a linear
relation between $\boldsymbol{\tau}$ and $\mathbf{C}_t$ leads to a non-linear relation between $\boldsymbol{\tau}$ and $\mathbf{C}_t^{-1}$.
It is only for small deformation that there is no ambiguity since the two for-
mulations are equivalent. This means that the form of Equation (11.48) is not
unique and that it would have been just as correct to write

$$\boldsymbol{\tau} = \int_0^\infty g(s)\left[\mathbf{C}_t^{-1}(t-s) - \mathbf{I}\right] ds \tag{11.49}$$

The choice between Equations (11.48), (11.49) or any other form must be
guided by experimental evidence. For example, in the case of a simple shear
flow, a particle located at $\mathbf{x}_t(x_{1t}, x_{2t}, x_{3t})$ at time t was at $\mathbf{x}(x_1, x_2, x_3)$ at time
t':

$$x_1 = x_{1t} + \dot{\gamma}\,x_{2t}(t' - t), \quad x_2 = x_{2t}, \quad x_3 = x_{3t}$$

We can thus compute the relative displacement gradient $\mathbf{F}_t$ and the dilatation
tensor $\mathbf{C}_t$

$$\mathbf{C}_t(t') = \begin{pmatrix} 1 & \dot{\gamma}(t'-t) & 0 \\ \dot{\gamma}(t'-t) & 1+\dot{\gamma}^2(t'-t)^2 & 0 \\ 0 & 0 & 1 \end{pmatrix} \tag{11.50}$$

The viscometric functions of Equation (11.48) are then

$$\mu_a = -\int_0^\infty s f(s)ds, \quad N_1 = -\dot{\gamma}^2\int_0^\infty s^2 f(s)ds, \quad N_2 = \dot{\gamma}^2\int_0^\infty s^2 f(s)ds$$

while those of Equation (11.49) are

$$\mu_a = \int_0^\infty sg(s)ds, \quad N_1 = \dot{\gamma}^2 \int_0^\infty s^2 g(s)ds, \quad N_2 = 0$$

We thus obtain normal stress differences in both cases. However, the experiments indicate that $|N_1|$ is always much larger than $|N_2|$. The law (11.48) thus does not predict the correct behaviour for the normal stress. In contrast, the law (11.49) leads to fairly realistic values of N_1 and N_2. As regards the apparent viscosity, it can depend on the shear rate if we assume that the memory function depends not only on time, but also on the deformations measured by the corresponding strain invariants. We can then write the viscoelastic law as

$$\boldsymbol{\sigma} = -p\mathbf{I} + \int_0^\infty g(s, J_{t2}, J_{t3}) \left[\mathbf{C}_t^{-1}(s) - \mathbf{I}\right] ds \tag{11.51}$$

where J_{t2} and J_{t3} are two invariants of tensor $\mathbf{C}_t(s)$. The third invariant $J_{t1} = \det\mathbf{C}$ is unity in view of the fluid incompressibility and can thus be eliminated. The apparent viscosity then depends on the shear rate through the invariants. Many other similar integral laws have been proposed in the literature. One must keep in mind, however, that the more complicated the law, the more difficult it is to determine its coefficient from experiments.

11.6.2 Non-Linear Differential Laws

We now seek a law that relates directly $\boldsymbol{\tau}$ and its derivatives to the deformation. Maxwell's law would be a good starting point, but it does not satisfy the objectivity principle! Indeed, the derivative $\dot{\boldsymbol{\tau}}$ is not an objective tensor (which is not a problem when we restrict ourselves to small deformations). An objective form of the time derivative of the stress is obtained when we compute it in a reference frame which has the translation velocity $\mathbf{u}$ of the fluid and rotates with the fluid vorticity $\boldsymbol{\omega}$. We thus define the 'Jaumann co-rotating derivative' which is objective:

$$\frac{\overset{\circ}{D}\boldsymbol{\tau}}{\overset{\circ}{D}t} = \frac{\partial\boldsymbol{\tau}}{\partial t} + \mathbf{u}\cdot\nabla\boldsymbol{\tau} + \boldsymbol{\tau}\cdot\boldsymbol{\omega} - \boldsymbol{\omega}\cdot\boldsymbol{\tau} \tag{11.52}$$

The simplest objective differential law is then the convected Maxwell model

$$\boldsymbol{\tau} + \lambda\frac{\overset{\circ}{D}\boldsymbol{\tau}}{\overset{\circ}{D}t} = 2\mu\mathbf{e} \tag{11.53}$$

or the convected Jeffrey law which is more symmetric:

$$\boldsymbol{\tau} + a_1\frac{\overset{\circ}{D}\boldsymbol{\tau}}{\overset{\circ}{D}t} = b_0\mathbf{e} + b_1\frac{\overset{\circ}{D}\mathbf{e}}{\overset{\circ}{D}t} \tag{11.54}$$

At steady state, the time dependency vanishes and we are back to the case of a purely viscous fluid (Section 11.3) with a non-linear behaviour due to the terms $\boldsymbol{\tau} \cdot \boldsymbol{\omega}$ and $\boldsymbol{\omega} \cdot \boldsymbol{\tau}$. For example, for a steady simple shear flow, one can check that the viscometric functions of Equation (11.53) are

$$\mu_a = \frac{\mu}{1 + \dot{\gamma}^2 \lambda^2}, \quad N_1 = \frac{2\dot{\gamma}^2 \mu}{1 + \dot{\gamma}^2 \lambda^2}, \quad N_2 = \frac{-\dot{\gamma}^2 \mu}{1 + \dot{\gamma}^2 \lambda^2}$$

We thus find that the apparent viscosity decreases with shear rate (as is often the case for polymer solutions) and that there are normal stress effects.

Other laws have been proposed in the literature. In general, it is preferable to use laws with only three parameters.

11.7   Non-Newtonian Flow Examples

We now enter the realm of non-Newtonian fluid mechanics and determine the flow of a fluid with a given constitutive law. This is done on a few examples which are simple enough to be amenable to an analytical solution.

11.7.1   Stationary 2D Flow between Two Parallel Plates

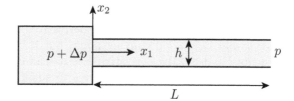

FIGURE 11.8
2D flow between two parallel flat plates separated by h.

A non-Newtonian fluid is driven by a pressure gradient through a rectangular section channel (height h, width w) with length L such that $h \ll w \ll L$. The configuration is thus similar to the Hele–Shaw cell (Chapter 4, Section 4.3.1). Such a device is used during extrusion of pasty products or to control the fluid viscosity on-line. In the first case, we know a priori the constitutive law and we want to compute the process control parameters such as the relation between the flow rate and the pressure drop (Examples (a) and (b)). In the second case, the analysis of the flow is used to determine the fluid rheological properties (Example (c)).

Far from the lateral walls ($x_3 = \pm w/2$), the flow is 2D, stationary and established. Thus the velocity has only one non-zero component $u_1(x_2)$ with

$u_1(\pm h/2) = 0$. The non-zero components of the shear rate tensor **e** are thus

$$e_{12} = e_{21} = \frac{1}{2}\frac{du_1(x_2)}{dx_2} = \frac{1}{2}\dot{\gamma}(x_2) \qquad (11.55)$$

For steady flow, the stress in the fluid is given by a general law of the type (11.15) and $\boldsymbol{\tau}$ depends only on $\dot{\gamma}(x_2)$. The general expression of the stress tensor in the fluid is then

$$\boldsymbol{\sigma} = \begin{pmatrix} \sigma_{11} & \sigma_{12} & 0 \\ \sigma_{12} & \sigma_{22} & 0 \\ 0 & 0 & \sigma_{33} \end{pmatrix} = \begin{pmatrix} -p + N_1 + \tau_{22} & \tau_{12} & 0 \\ \tau_{12} & -p + \tau_{22} & 0 \\ 0 & 0 & -p + \tau_{22} - N_2 \end{pmatrix}$$

where we have introduced the two normal stress differences N_1 and N_2 which depend only on $\dot{\gamma}(x_2)$. Since inertia forces are identically zero for this uniaxial flow, the flow equations are simply

$$-\frac{\partial p}{\partial x_1} + \frac{\partial \tau_{12}}{\partial x_2} = 0, \qquad -\frac{\partial p}{\partial x_2} + \frac{\partial \tau_{22}}{\partial x_2} = 0, \qquad -\frac{\partial p}{\partial x_3} = 0$$

Taking the derivative of the three equations with respect to x_1 and adding, we find that p depends only on x_1 and that its gradient is constant (just like for the flow of a Newtonian fluid). We note

$$-\frac{\partial p}{\partial x_1} = A = \Delta P/L$$

where A is the pressure drop per unit length and ΔP is the pressure drop over the channel length L (note that this pressure drop was denoted G in Chapter 4; the notation is changed here to avoid a confusion with the fluid elastic modulus). We thus obtain the shear stress in the fluid:

$$\tau_{12} = -A\,x_2 \qquad (11.56)$$

where we have used the fact that τ_{12} is an odd function of x_2.

11.7.1.1 Example (a): Power Law Fluid

If we know the fluid constitutive law, we only have to replace τ_{12} by its expression in terms of e_{12} in Equation (11.56) and proceed with the solution to obtain the velocity field in the channel. For example, in the case of a fluid that satisfies the power law (11.17), the shear rate is given by

$$\dot{\gamma}(x_2) = \frac{\tau_{12}}{k}\left[\frac{|\tau_{12}|}{k}\right]^{(\frac{1}{n}-1)}$$

or

$$\frac{du_1(x_2)}{dx_2} = -\left(\frac{A}{k}\right)^{\frac{1}{n}} x_2\,|x_2|^{(\frac{1}{n}-1)}$$

We integrate to obtain the velocity field between the walls:

$$u_1(x_2) = -\left(\frac{A}{k}\right)^{\frac{1}{n}} \frac{n}{n+1} \left[|x_2|^{\frac{1}{n}+1} - \left(\frac{h}{2}\right)^{\frac{1}{n}+1} \right]$$

and the flow rate per unit width:

$$Q = \int_{-h/2}^{h/2} u_1\, dx_2 = 2\left(\frac{A}{k}\right)^{\frac{1}{n}} \frac{n}{2n+1} \left(\frac{h}{2}\right)^{\frac{1}{n}+2}$$

Note that for $n = 1$, we recover the expression of the flow rate of a Newtonian fluid in a Hele–Shaw cell (Chapter 4, Section 4.3.1). The mean velocity $\bar{u}$ is then

$$\bar{u} = \frac{Q}{h} = \left(\frac{A}{k}\right)^{\frac{1}{n}} \frac{n}{2n+1} \left(\frac{h}{2}\right)^{\frac{1}{n}+1}$$

We can then write the flow field in the form

$$u_1(x_2) = \bar{u}\frac{1+2n}{n+1}\left[1 - \left(\frac{|x_2|}{h/2}\right)^{\frac{1}{n}+1}\right]$$

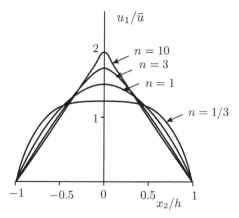

FIGURE 11.9
Velocity profiles between two parallel plates for a power law fluid for different values of the flow index n.

The velocity profiles obtained for different values of the flow index n are shown in Figure 11.9. We note a flattening of the profile at the channel centre ($x_2 = 0$) for $n < 1$. This is due to the fact that $\dot{\gamma} \to 0$ on the axis and thus $\mu_a \to \infty$, which leads to plug flow in the channel centre. When $n > 1$, the velocity profile has a sharp maximum on the axis. This is due to the fact that $\mu_a \to 0$ for $x_2 = 0$, which leads to a tendency towards slip between the two fluid layers $x_2 > 0$ and $x_2 < 0$.

11.7.1.2 Example (b): Bingham Fluid

Now let us consider another case where the fluid has a yield stress and obeys the Bingham constitutive equation (11.20). Equation (11.56) implies that the fluid is sheared for $|x_2| \geq b$ where $b = \tau_s/A$. In the centre of the channel, the fluid is not sheared but moves as a whole with constant velocity u_b (plug flow). Bingham's law leads to the following relations for $x_2 \geq 0$:

$$\dot\gamma = 0 \quad \text{if} \quad 0 \leq x_2 \leq b \tag{11.57}$$

$$\dot\gamma = \frac{\tau_{12} + \tau_s}{\mu_p} = \frac{-Ax_2 + \tau_s}{\mu_p} \quad \text{if} \quad b \leq x_2 \leq h/2 \tag{11.58}$$

where we have used the fact that the shear stress was negative. For $x_2 \leq 0$, we have the symmetric relations $\dot\gamma = (-Ax_2 - \tau_s)/\mu_p$ for $-h/2 \leq x_2 \leq -b$. Integrating (11.58) with condition $u_1(h/2) = 0$ leads us to

$$u_1 = -\frac{A}{2\mu_p}(x_2^2 - h^2/4) + \frac{\tau_s}{\mu_p}(x_2 - h/2) \tag{11.59}$$

We thus can compute the velocity u_b of the central plug

$$u_b = u_1(b) = -\frac{A}{2\mu_p}(b^2 - h^2/4) + \frac{\tau_s}{\mu_p}(b - h/2) \tag{11.60}$$

As shown in Figure 11.10, the velocity profile is flat in the channel centre (plug flow) and parabolic near the walls. The flow rate per unit width is then

$$Q = 2\left\{ bu_b + \int_b^{h/2} \left[-\frac{A}{2\mu_p}(x_2^2 - h^2/4) + \frac{\tau_s}{\mu_p}(x_2 - h/2) \right] dx_2 \right\} \tag{11.61}$$

which leads to

$$Q = -\frac{2A}{3\mu_p}(b^3 - h^3/8) + \frac{\tau_s}{\mu_p}(b^2 - h^2/4)$$

We can verify that for a Newtonian fluid for which τ_s and b are zero, we do retrieve the result obtained for a Hele–Shaw flow (Chapter 4, Section 4.3.1).

Of course, if the pressure gradient A is too small, the fluid will not flow. The minimum value of A for flow to occur corresponds to $b < h/2$, or $A > 2\tau_s/h$. Note that it is sometimes desirable to prevent flow from happening under normal use: tooth paste (a yield stress fluid) does not flow when you open the tube, but does flow when you apply a pressure drop larger than gravity by squeezing the tube.

In conclusion, in configurations where the shear rate varies locally, the flow of a non-Newtonian fluid becomes quite complex.

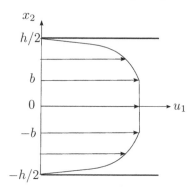

FIGURE 11.10
Velocity profile of a yield stress fluid flowing between two flat plates: at the
centre of the channel where the shear stress is lower than the yield stress, the
fluid undergoes plug flow; near the walls, the fluid is sheared and the velocity
profile is parabolic.

11.7.1.3 Example (c): Unknown Constitutive Law

We finally consider the case where we do not know the constitutive law
$\tau_{12} = \tau_{12}[\dot{\gamma}(x_2)]$ of the fluid and wish to determine it from the simultane-
ous measurement of the pressure drop $\Delta P/L$ and of the flow rate Q under
different flow conditions.

We first invert the constitutive law and write it as

$$\dot{\gamma}(x_2) = f(\tau_{12}) \tag{11.62}$$

where f is an unknown function to be determined. The velocity between the
two walls is thus

$$u_1(x_2) = -\int_{-h/2}^{x_2} f(Ax_2)dx_2 \tag{11.63}$$

The flow rate Q per unit width can be integrated by parts:

$$Q = \int_{-h/2}^{h/2} u_1\,dx_2 = -\int_{-h/2}^{h/2} x_2\,du_1 = -\int_{-h/2}^{h/2} x_2\dot{\gamma}dx_2 \tag{11.64}$$

When we introduce the inverted constitutive law, we find

$$Q = \int_{-h/2}^{h/2} x_2 f(Ax_2)dx_2 = \frac{2}{A^2}\int_0^{Ah/2}\zeta f(\zeta)d\zeta \tag{11.65}$$

We now compute the derivative of the flow rate with respect to the pressure
gradient:

$$\frac{\partial(A^2 Q)}{\partial A} = \frac{Ah^2}{2}f(Ah/2) \tag{11.66}$$

The identification technique then consists of measuring simultaneously Q and $A = \Delta P/L$ for different values of Q (or of ΔP, depending on how the device is controlled), of computing numerically the derivative $\partial(A^2 Q)/\partial A$ and thus determining the function $f(\zeta)$ point by point.

This technique can also be used with cylindrical capillary tubes. We assume that the tube length is large enough with respect to the radius R for entrance and end effects to be negligible. We then measure the flow rate Q as a function of the pressure gradient A which are related by

$$\frac{\partial(A^3 Q)}{\partial A} = -\pi A^2 R^3 f(AR) \tag{11.67}$$

This relation, which is equivalent to Equation (11.66), is known as the Mooney–Rabinovich equation.

11.7.2 Oscillatory Flow of a Viscoelastic Liquid

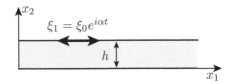

FIGURE 11.11
Oscillatory flow between two parallel flat plates.

We now turn to the case where the fluid is still between two parallel flat plates, but the the flow is not created by a pressure gradient, but by the oscillation of the upper plate $x_2 = h$ while the lower plate $x_2 = 0$ is fixed. The displacement of the upper plate is given by

$$\xi_1 = \xi_0 e^{i\alpha t}, \quad \xi_2 = \xi_3 = 0 \quad \text{at} \quad x_2 = h \tag{11.68}$$

where ξ_0 is the complex displacement amplitude and α is the angular frequency. The velocity boundary conditions on each plate are

$$u_1 = u_2 = u_3 = 0 \quad \text{at} \quad x_2 = 0$$

$$u_1 = i\alpha\xi_0 e^{i\alpha t}, \quad u_2 = u_3 = 0 \quad \text{at} \quad x_2 = h$$

The velocity field

$$u_1 = i\alpha\xi_0 e^{i\alpha t}\left(\frac{x_2}{h}\right), \quad u_2 = u_3 = 0 \tag{11.69}$$

satisfies the boundary conditions. The only non-zero components of the shear rate tensor are constant in the fluid domain

$$e_{12} = e_{21} = \frac{1}{2h} i\alpha\xi_0 e^{i\alpha t} \tag{11.70}$$

To this homogeneous shear field corresponds an homogeneous stress field given by

$$\tau_{12} = \tau_{21} = \tau_0 e^{i\alpha t} \tag{11.71}$$

which thus satisfies the flow equations of motion with no inertia.

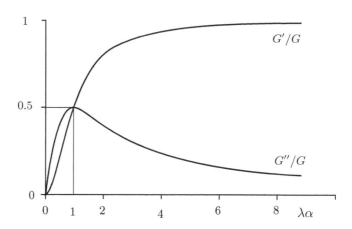

FIGURE 11.12
Conservation G' and loss G'' moduli for a Maxwell fluid.

We now wish to find the relation between the complex amplitudes of the stress τ_0 and of the displacement ξ_0. We assume that the fluid obeys the Maxwell law (11.33)

$$\tau_{12} + \lambda \frac{d\tau_{12}}{dt} = 2\mu e_{12} = \mu i \alpha \left(\frac{\xi_0}{h}\right) e^{i\alpha t} \tag{11.72}$$

When we replace τ_{12} by its value, we find

$$\frac{\tau_0}{\xi_0/h} = \frac{i\mu\alpha}{1 + i\lambda\alpha} = \frac{\lambda\mu\alpha^2}{1 + \lambda^2\alpha^2} + i\frac{\mu\alpha}{1 + \lambda^2\alpha^2} \tag{11.73}$$

The complex shear modulus G^* is defined by

$$G^* = \frac{\tau_0}{\xi_0/h} \tag{11.74}$$

which can be decomposed into the real G' and imaginary G'' parts,

$$G^* = G'(\alpha) + iG''(\alpha) \tag{11.75}$$

We can write the real and imaginary parts of G^* in terms of the elastic modulus $G = \mu/\lambda$ and the viscosity μ (which are both real!) of Maxwell's law

$$G' = \frac{G\mu^2\alpha^2}{G^2 + \mu^2\alpha^2} \quad \text{and} \quad G'' = \frac{\mu G^2\alpha}{G^2 + \mu^2\alpha^2} \tag{11.76}$$

or

$$\frac{G'}{G} = \frac{\lambda^2 \alpha^2}{1 + \lambda^2 \alpha^2} \quad \text{and} \quad \frac{G''}{G} = \frac{\lambda \alpha}{1 + \lambda^2 \alpha^2} \tag{11.77}$$

The frequency evolution of G' and G'' for a Maxwell fluid is shown in Figure 11.12. We note that for a purely elastic fluid ($\mu \to \infty$), we find $G' = G$ and $G'' = 0$, while for a purely viscous fluid ($G \to \infty$) we find $G' = 0$ and $G'' = \mu \alpha$. The energy storage property of the fluid is thus measured by G', which is called the 'storage modulus'. Conversely, the viscous dissipation in the fluid is measured by G'', which is called the 'loss modulus'. The two moduli G' and G'' can be measured by means of a rotative viscometer (like those shown in Figure 11.1) where we impose a sinusoidal oscillation.

11.8 Conclusion

Rheology and non-Newtonian fluid mechanics are two very active research domains with a wide range of applications in nature and in industry. We should keep in mind that the constitutive laws are empirical and have a restricted domain of validity. They can be used to interpolate experimental data over a given range of parameters (e.g. shear rate). It is dangerous to extrapolate them far from their limits of determination. Fluids with a time-dependent behaviour present some very specific non-linear flow properties. Non-Newtonian fluid mechanics thus offer a wide variety of fascinating problems More about non-Newtonian fluids can be found in the books by Tanner [52] or Coussot [17] among others.

11.9   Problems

11.9.1   Flow of a Bingham Fluid in a Cylindrical Tube

From an Ecole Polytechnique problem written with E. Guazzelli

The rheological behaviour of some heavy oils can be approximated by a Bingham law (viscosity μ_p, yield stress τ_s). The objective of the problem is to compute the flow of such an oil in a cylindrical tube with radius R under a pressure difference Δp over a tube length L. We use cylindrical coordinates centred on the tube axis.

1. We assume that the flow is axisymmetric and that $\sigma_{zz} = -p$. Write the axial component of the momentum equation.

2. Find the shear stress τ_{zr} in terms of the radial position r and of the wall shear stress $\tau_p = \tau_{zr}(R)$.

3. Show that the flow rate is given by

$$Q = -\pi \int_{u_z(0)}^{u_z(R)} r^2 du_z$$

4. Compute Q in terms of τ_{zr} and of the shear rate $\dot{\gamma} = du_z/dr$.

5. Show that for a Bingham fluid, the flow rate is given by

$$Q = -\frac{\pi R^3}{\mu_p \tau_p^3} \int_{\tau_s}^{\tau_p} \tau_{zr}^2 (\tau_{zr} - \tau_s) d\tau_{zr}$$

6. We introduce the flow rate Q_P of a Newtonian fluid under the same pressure drop (Poiseuille law). Show that the flow rate of a Bingham fluid can be written as

$$Q/Q_P = 1 - \frac{4\tau_s}{3\tau_p} + \frac{\tau_s^4}{3\tau_p^4}$$

7. Show that for a given pressure drop, the fluid will flow only in tubes with a radius larger than some critical value to be determined.

8. The flow of crude oil in a porous medium such as a drilling well can be approximately modelled as the flow through parallel cylindrical pores with radius R. What information on the pore size distribution can be obtained from the experimental relation between the flow rate and the pressure drop?

9. Assume that oil ($\tau_s = 8.67$ Pa, $\mu_p = 0.245$ Pa s) flows through a 10 cm thick slab of porous medium under a pressure drop 7×10^5 Pa. What is the minimum pore diameter through which the oil can flow?

11.9.2 Weissenberg Effect

From an Ecole Polytechnique problem written with E. Guazzelli

The effect of normal stress differences is observed in some high-molecular-weight polymer solutions. To put it into evidence, we plunge a vertical cylindrical rod (radius a) in a non-Newtonian fluid (density ρ) and rotate it with angular velocity ω. We then observe that the free surface of the liquid rises along the rod (Figure 11.13): this is the so-called 'Weissenberg' effect.

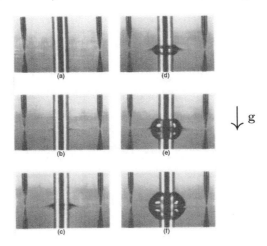

FIGURE 11.13
Rise of a polyisobutylene solution in oil along a vertical rotating rod of radius 4.76 mm. The rotation speed (rotation per second) is (a) 0; (b) 1.0; (c) 2.2; (d) 3.0; (e) 5.0; (f) 7.0.(Reproduced from Beavers an Joseph [9], with permission from Cambridge University Press.)

The objective of the problem is to analyse this effect in a simplified situation. We use cylindrical coordinates where the z-axis is along the rod axis and is oriented upwards. The free surface of the liquid is $z = h(r, \omega)$. The following assumptions are made:

- The atmospheric pressure is the reference pressure, thus outside the liquid $p = 0$
- Surface tension effects between air and the liquid are negligible
- The liquid surface is initially planar with equation $z = 0$ (the contact angle between the rod and the liquid is 90° before the rod is rotated)
- The liquid surface deformation is small ($|h|/a \ll 1$ and $|\partial h / \partial r| \ll 1$)
- The flow is a steady rotation which depends only on r:

$$\mathbf{u} = u_\theta(r)\mathbf{e}_\theta$$

- The fluid has a non-Newtonian behaviour characterised by

$$\tau = 2\mu \mathbf{e}$$

$$N_1 = \sigma_{11} - \sigma_{22} = B_1 \dot{\gamma}^2, \qquad N_2 = \sigma_{22} - \sigma_{33} = B_2 \dot{\gamma}^2$$

where $\mathbf{e}$ is the shear rate tensor, $\dot{\gamma}$ is the shear rate, μ is the constant shear viscosity, N_1 and N_2 are the normal stress difference in a simple shear flow ($\mathbf{u} = \dot{\gamma} x_2 \mathbf{e}_1$), and B_1 and B_2 are two constant coefficients

1. Find the non-zero components of the shear rate $\mathbf{e}$ and stress σ tensors.

2. Write the equation of motion of the fluid without assuming that inertia and gravity are negligible.

3. Find the velocity boundary conditions.

4. Determine the velocity field $u_\theta(r)$ and deduce the shear rate $\dot{\gamma}(r)$.

5. Show that the normal stress σ_{zz} is hydrostatic.

6. Compute $\sigma_{rr}(r, z)$ in terms of a, ω, ρ and B_1.

7. Find the equation of the free surface $h(r, \omega)$.

8. Show that the height of the fluid on the rod $h(a, \omega)$ is given by

$$h(a, \omega) = \frac{\omega^2}{\rho g} \left[(B_1 + 4B_2) - \frac{\rho a^2}{2} \right]$$

Under which conditions does the fluid climb on the rod? What would happen if the fluid were Newtonian?

9. In the experimental study of Beavers and Joseph [9], the fluid was a polyisobutylene solution in a petroleum oil, which was 15 000 times more viscous than water (for a shear rate between 1 and 10 s^{-1}) and had a density of 0.89 g/cm^3. The rod surface was hydrophobic to ensure the flatness of the liquid surface, as can be observed in Figure 11.13. Thus surface tension effects were negligible as assumed in the model. The height of the liquid was measured as a function of the diameter a and of the rod rotation speed for $\omega \leq 10 \sim 16$ s^{-1}. It was found

$$h(a) \approx 8 \times 10^{-4} \omega^2 \quad \text{for} \quad a = 3.17 \text{ mm} \quad \text{and} \quad a = 4.76 \text{ mm}$$

$$h(a) \approx 10^{-3} \omega^2 \quad \text{for} \quad a = 6.35 \text{ mm}$$

(a) Compare the model to the experimental results for the two small rods ($a = 3.17$ mm and $a = 4.76$ mm).

(b) Find the value of $B_1 + 4B_2$. Comment on your result.

(c) For what maximum rod radius will the fluid climb?

(d) Is it correct to neglect inertia effects for the largest rod ($a = 6.35$ mm)? Compare the influence of a as computed and as observed experimentally. Comment.

Appendix A

Notations

CONTENTS

A.1   Vectors and Tensors

We use a Cartesian reference frame with origin O and base vectors $(\mathbf{e}_1, \mathbf{e}_2, \mathbf{e}_3)$. A point M in space is defined by either

- The position vector $\mathbf{OM} = \mathbf{x}$

- The components x_1, x_2, x_3 of $\mathbf{x}$, which are noted simply x_i where $i = 1, 2, 3$

- A column matrix
$$\mathbf{x} = \begin{pmatrix} x_1 \\ x_2 \\ x_3 \end{pmatrix}$$

Similarly, a second order tensor $\mathbf{C}$ is denoted by either

- A bold face symbol $\mathbf{C}$

- The components C_{ij} where $i = 1, 2, 3$ and $j = 1, 2, 3$

- A 3×3 matrix
$$\mathbf{C} = \begin{pmatrix} C_{11} & C_{12} & C_{13} \\ C_{23} & C_{22} & C_{23} \\ C_{31} & C_{32} & C_{33} \end{pmatrix}$$

The gradient of a scalar function p is then denoted

$$\nabla p = \partial p / \partial x_i$$

A.2   Einstein Summation Convention

Einstein summation convention is that we sum automatically from 1 to 3 on any repeated index and, correspondlingly, we omit the summation symbol Σ. For example, the scalar product between two vectors **u** and **v**, which should be written as

$$\mathbf{u} \cdot \mathbf{v} = u_1 v_1 + u_2 v_2 + u_3 v_3 = \sum_{i=1}^{3} u_i v_i$$

with Einstein convention, becomes

$$\mathbf{u} \cdot \mathbf{v} = u_i v_i$$

The divergence of a vector **v** or of a tensor $\boldsymbol{\sigma}$ is then

$$\nabla \cdot \mathbf{v} = \partial v_i / \partial x_i, \qquad \nabla \cdot \boldsymbol{\sigma} = \partial \sigma_{ij} / \partial x_i$$

The dot product of a tensor $\boldsymbol{\sigma}$ and a vector $\boldsymbol{n}$ is

$$\boldsymbol{\sigma} \cdot \mathbf{n} = \sigma_{ij} n_j$$

which also corresponds to the product of a 3×3 matrix by a column vector

$$\boldsymbol{\sigma} \cdot \mathbf{n} = \begin{pmatrix} \sigma_{11} & \sigma_{12} & \sigma_{13} \\ \sigma_{23} & \sigma_{22} & \sigma_{23} \\ \sigma_{31} & \sigma_{32} & \sigma_{33} \end{pmatrix} \begin{pmatrix} n_1 \\ n_2 \\ n_3 \end{pmatrix}$$

The Kronecker symbol δ_{ij} is the index representation of the unit 3×3 matrix

$$\begin{aligned} \delta_{ij} &= 1 \quad \text{if} \quad i = j \\ &= 0 \quad \text{if} \quad i \neq j \end{aligned}$$

The permutation symbol is defined by

$$\begin{aligned} \varepsilon_{ijk} &= 1 \quad \text{if } i, j, k \text{ are a permutation of } 1, 2, 3 \\ &= -1 \quad \text{if } i, j, k \text{ are an anti-permutation of } 1, 2, 3 \\ &= 0 \quad \text{if two indices are repeated} \end{aligned}$$

The permutation symbol is used to write the cross product of two vectors **u** and **v**

$$\mathbf{u} \times \mathbf{v} = \varepsilon_{ijk} u_j v_k$$

Similarly, the curl of a vector **u** is

$$\text{curl } \mathbf{u} = \nabla \times \mathbf{u} = \varepsilon_{ijk} \frac{\partial u_k}{\partial x_j}$$

A.3 Integration on a Sphere

Let S be a sphere centred on O with radius a and outer unit normal vector $\mathbf{n}$. The solid angle centred on O is denoted Ω. The integrals of the products of the components of $\mathbf{n}$ on the sphere are given by

$$\int_S n_i n_j d\Omega = \frac{4\pi}{3} \delta_{ij}$$

$$\int_S n_i n_j n_k d\Omega = 0$$

$$\int_S n_i n_j n_k n_m d\Omega = \frac{4\pi}{15} \left(\delta_{ij}\delta_{km} + \delta_{im}\delta_{kj} + \delta_{ik}\delta_{jm} \right)$$

$$\int_S n_i n_j n_k n_a n_b d\Omega = 0$$

$$\int_S n_i n_j n_k n_m n_a n_b d\Omega = \frac{4\pi}{105} \left(\delta_{ij}\delta_{km}\delta_{ab} + \ldots \right)$$

Appendix B

Curvilinear Coordinates

CONTENTS

B.1   Cylindrical Coordinates

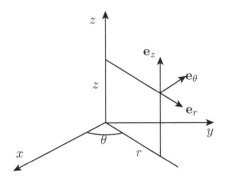

FIGURE B.1
Cylindrical coordinates $\{r, \theta, z\}$.

Position of a point

$$\mathbf{x} = r\mathbf{e}_r + z\mathbf{e}_z$$

$$d\mathbf{x} = dr\mathbf{e}_r + rd\theta\mathbf{e}_\theta + dz\mathbf{e}_z$$

Gradient of a scalar function p

$$\nabla p = \frac{\partial p}{\partial r}\mathbf{e}_r + \frac{1}{r}\frac{\partial p}{\partial \theta}\mathbf{e}_\theta + \frac{\partial p}{\partial z}\mathbf{e}_z$$

Laplacian of a scalar function p

$$\Delta p = \nabla^2 p = \frac{\partial^2 p}{\partial r^2} + \frac{1}{r}\frac{\partial p}{\partial r} + \frac{1}{r^2}\frac{\partial^2 p}{\partial \theta^2} + \frac{\partial^2 p}{\partial z^2}$$

Divergence of a vector $\mathbf{v}$

$$\nabla \cdot \mathbf{v} = \frac{\partial v_r}{\partial r} + \frac{v_r}{r} + \frac{1}{r}\frac{\partial v_\theta}{\partial \theta} + \frac{\partial v_z}{\partial z}$$

Curl of a vector $\mathbf{v}$

$$\text{curl } \mathbf{v} = \nabla \times \mathbf{v} = \begin{pmatrix} \frac{1}{r}\frac{\partial v_z}{\partial \theta} - \frac{\partial v_\theta}{\partial z} \\[2mm] \frac{\partial v_r}{\partial z} - \frac{\partial v_z}{\partial r} \\[2mm] \frac{1}{r}\frac{\partial}{\partial r}\left(r v_\theta\right) - \frac{1}{r}\frac{\partial v_r}{\partial \theta} \end{pmatrix}$$

Laplacian of a vector $\mathbf{v}$

$$\Delta \mathbf{v} = \nabla^2 \mathbf{v} = \begin{pmatrix} \frac{\partial}{\partial r}\left(\frac{1}{r}\frac{\partial}{\partial r}\left(r v_r\right)\right) + \frac{1}{r^2}\frac{\partial^2 v_r}{\partial \theta^2} - \frac{2}{r^2}\frac{\partial v_\theta}{\partial \theta} + \frac{\partial^2 v_r}{\partial z^2} \\[2mm] \frac{\partial}{\partial r}\left(\frac{1}{r}\frac{\partial}{\partial r}\left(r v_\theta\right)\right) + \frac{1}{r^2}\frac{\partial^2 v_\theta}{\partial \theta^2} + \frac{2}{r^2}\frac{\partial v_r}{\partial \theta} + \frac{\partial^2 v_\theta}{\partial z^2} \\[2mm] \frac{1}{r}\frac{\partial}{\partial r}\left(r \frac{\partial}{\partial r} v_z\right) + \frac{1}{r^2}\frac{\partial^2 v_z}{\partial \theta^2} + \frac{\partial^2 v_z}{\partial z^2} \end{pmatrix}$$

Rate of strain tensor

$$\mathbf{e} = \frac{1}{2}\begin{pmatrix} 2\frac{\partial u_r}{\partial r} & \frac{1}{r}\left(\frac{\partial u_r}{\partial \theta} - u_\theta\right) + \frac{\partial u_\theta}{\partial r} & \frac{\partial u_z}{\partial r} + \frac{\partial u_r}{\partial z} \\[2mm] \frac{1}{r}\left(\frac{\partial u_r}{\partial \theta} - u_\theta\right) + \frac{\partial u_\theta}{\partial r} & 2\frac{1}{r}\left(\frac{\partial u_\theta}{\partial \theta} + u_r\right) & \frac{1}{r}\frac{\partial u_z}{\partial \theta} + \frac{\partial u_\theta}{\partial z} \\[2mm] \frac{\partial u_z}{\partial r} + \frac{\partial u_r}{\partial z} & \frac{1}{r}\frac{\partial u_z}{\partial \theta} + \frac{\partial u_\theta}{\partial z} & 2\frac{\partial u_z}{\partial z} \end{pmatrix}$$

Divergence of a symmetric tensor $\boldsymbol{\sigma}$

$$\nabla \cdot \sigma = \begin{pmatrix} \frac{1}{r}\frac{\partial}{\partial r}\left(r\sigma_{rr}\right) + \frac{1}{r}\frac{\partial}{\partial \theta}\sigma_{r\theta} - \frac{1}{r}\sigma_{\theta\theta} + \frac{\partial}{\partial z}\sigma_{rz} \\[2mm] \frac{1}{r}\frac{\partial}{\partial \theta}\sigma_{\theta\theta} + \frac{\partial}{\partial r}\sigma_{r\theta} + \frac{2}{r}\sigma_{r\theta} + \frac{\partial}{\partial z}\sigma_{\theta z} \\[2mm] \frac{1}{r}\frac{\partial}{\partial r}\left(r\sigma_{rz}\right) + \frac{1}{r}\frac{\partial}{\partial \theta}\sigma_{\theta z} + \frac{\partial}{\partial z}\sigma_{zz} \end{pmatrix}$$

Convective derivative of a vector $\mathbf{u}$

$$\frac{D\mathbf{u}}{Dt} = \begin{pmatrix} \frac{\partial u_r}{\partial t} + u_r\frac{\partial u_r}{\partial r} + \frac{u_\theta}{r}\frac{\partial u_r}{\partial \theta} - \frac{u_\theta^2}{r} + u_z\frac{\partial u_r}{\partial z} \\[2mm] \frac{\partial u_\theta}{\partial t} + u_r\frac{\partial u_\theta}{\partial r} + \frac{u_\theta}{r}\frac{\partial u_\theta}{\partial \theta} + \frac{u_r u_\theta}{r} + u_z\frac{\partial u_\theta}{\partial z} \\[2mm] \frac{\partial u_z}{\partial t} + u_r\frac{\partial u_z}{\partial r} + \frac{u_\theta}{r}\frac{\partial u_z}{\partial \theta} + u_z\frac{\partial u_z}{\partial z} \end{pmatrix}$$

Stokes equations

$$\frac{\partial u_r}{\partial r} + \frac{u_r}{r} + \frac{1}{r}\frac{\partial u_\theta}{\partial \theta} + \frac{\partial u_z}{\partial z} = 0$$

$$-\frac{\partial p}{\partial r} + \mu \left[\frac{\partial}{\partial r}\left(\frac{1}{r}\frac{\partial}{\partial r}(ru_r)\right) + \frac{1}{r^2}\frac{\partial^2 u_r}{\partial \theta^2} - \frac{2}{r^2}\frac{\partial u_\theta}{\partial \theta} + \frac{\partial^2 u_r}{\partial z^2}\right] = 0$$

$$-\frac{\partial p}{r\partial \theta} + \mu \left[\frac{\partial}{\partial r}\left(\frac{1}{r}\frac{\partial}{\partial r}(ru_\theta)\right) + \frac{1}{r^2}\frac{\partial^2 u_\theta}{\partial \theta^2} + \frac{2}{r^2}\frac{\partial u_r}{\partial \theta} + \frac{\partial^2 u_\theta}{\partial z^2}\right] = 0$$

$$-\frac{\partial p}{\partial z} + \mu \left[\frac{1}{r}\frac{\partial}{\partial r}\left(r\frac{\partial}{\partial r}u_z\right) + \frac{1}{r^2}\frac{\partial^2 u_z}{\partial \theta^2} + \frac{\partial^2 u_z}{\partial z^2}\right] = 0$$

Stream function Ψ

$$u_r = \frac{1}{r}\frac{\partial \Psi}{\partial \theta}, \qquad u_\theta = -\frac{\partial \Psi}{\partial r}$$

Stokes operator $\mathfrak{D}^2$

$$\mathfrak{D}^2\Psi = \frac{1}{r}\frac{\partial}{\partial r}\left(r\frac{\partial \Psi}{\partial r}\right) + \frac{1}{r^2}\frac{\partial^2 \Psi}{\partial \theta^2}$$

B.2   Spherical Coordinates

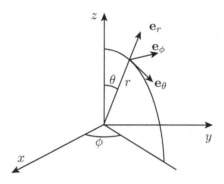

FIGURE B.2
Spherical coordinates $\{r, \theta, \phi\}$.

Position of a point

$$\mathbf{x} = r\mathbf{e}_r$$

$$\mathbf{dx} = dr\mathbf{e}_r + rd\theta\mathbf{e}_\theta + r\sin\theta d\phi\mathbf{e}_\phi$$

Gradient of a scalar function p

$$\nabla p = \frac{\partial p}{\partial r}\mathbf{e}_r + \frac{1}{r}\frac{\partial p}{\partial \theta}\mathbf{e}_\theta + \frac{1}{r\sin\theta}\frac{\partial p}{\partial \phi}\mathbf{e}_\phi$$

Laplacian of a scalar function p

$$\Delta p = \nabla^2 p = \frac{\partial^2 p}{\partial r^2} + \frac{2}{r}\frac{\partial p}{\partial r} + \frac{1}{r^2}\frac{\partial^2 p}{\partial \theta^2} + \frac{1}{r^2}\cot\theta\frac{\partial p}{\partial \theta} + \frac{1}{r^2\sin^2\theta}\frac{\partial^2 p}{\partial \phi^2}$$

Divergence of a vector $\mathbf{v}$

$$\nabla \cdot \mathbf{v} = \frac{\partial v_r}{\partial r} + 2\frac{v_r}{r} + \frac{1}{r}\frac{\partial v_\theta}{\partial \theta} + \frac{1}{r\sin\theta}\frac{\partial v_\phi}{\partial \phi} + \cot\theta\frac{v_\theta}{r}$$

Curl of a vector $\mathbf{v}$

$$\text{curl } \mathbf{v} = \nabla \times \mathbf{v} = \begin{pmatrix} \frac{1}{r\sin\theta}\left(\frac{\partial}{\partial \theta}(v_\phi\sin\theta) - \frac{\partial v_\theta}{\partial \phi}\right) \\[2ex] \frac{1}{r\sin\theta}\frac{\partial v_r}{\partial \phi} - \frac{1}{r}\frac{\partial}{\partial r}(rv_\phi) \\[2ex] \frac{1}{r}\frac{\partial}{\partial r}(rv_\theta) - \frac{1}{r}\frac{\partial v_r}{\partial \theta} \end{pmatrix}$$

Laplacian of a vector **v**

$$\Delta \mathbf{v} = \nabla^2 \mathbf{v} = \begin{pmatrix} \Delta v_r - \frac{2v_r}{r^2} - \frac{2}{r^2}\frac{\partial v_\theta}{\partial \theta} - \frac{2v_\theta \cot \theta}{r^2} - \frac{2}{r^2 \sin \theta}\frac{\partial v_\phi}{\partial \phi} \\[2mm] \Delta v_\theta + \frac{2}{r^2}\frac{\partial v_r}{\partial \theta} - \frac{v_\theta}{r^2 \sin^2 \theta} - \frac{2\cos \theta}{r^2 \sin^2 \theta}\frac{\partial v_\phi}{\partial \phi} \\[2mm] \Delta v_\phi - \frac{v_\phi}{r^2 \sin^2 \theta} + \frac{2}{r^2 \sin \theta}\frac{\partial v_r}{\partial \phi} + \frac{2\cos \theta}{r^2 \sin^2 \theta}\frac{\partial v_\theta}{\partial \phi} \end{pmatrix}$$

Rate of strain tensor

$$e_{rr} = \frac{\partial u_r}{\partial r}, \qquad e_{\theta\theta} = \frac{1}{r}\left(\frac{\partial u_\theta}{\partial \theta} + u_r\right)$$

$$e_{\phi\phi} = \frac{1}{r\sin\theta}\frac{\partial u_\phi}{\partial \phi} + \frac{\cot\theta}{r}u_\theta + \frac{u_r}{r}, \qquad e_{r\theta} = \frac{1}{2}\left(\frac{\partial u_\theta}{\partial r} + \frac{1}{r}\left(\frac{\partial u_r}{\partial \theta} - u_\theta\right)\right)$$

$$e_{r\phi} = \frac{1}{2}\left(\frac{\partial u_\phi}{\partial r} + \frac{1}{r\sin\theta}\frac{\partial u_r}{\partial \phi} - \frac{u_\phi}{r}\right), \qquad e_{\phi\theta} = \frac{1}{2}\left(\frac{1}{r}\frac{\partial u_\phi}{\partial \theta} + \frac{1}{r\sin\theta}\frac{\partial u_\theta}{\partial \phi} - \frac{\cot\theta}{r}u_\phi\right)$$

Divergence of a symmetric tensor $\boldsymbol{\sigma}$

$$\nabla \cdot \boldsymbol{\sigma} = \begin{pmatrix} \frac{1}{r^2}\frac{\partial}{\partial r}(r^2\sigma_{rr}) + \frac{1}{r\sin\theta}\frac{\partial}{\partial \theta}(\sin\theta\,\sigma_{r\theta}) + \frac{1}{r\sin\theta}\frac{\partial \sigma_{r\phi}}{\partial \phi} - \frac{\sigma_{\theta\theta}+\sigma_{\phi\phi}}{r} \\[2mm] \frac{1}{r^2}\frac{\partial}{\partial r}(r^2\sigma_{r\theta}) + \frac{1}{r\sin\theta}\frac{\partial}{\partial \theta}(\sin\theta\,\sigma_{\theta\theta}) + \frac{1}{r\sin\theta}\frac{\partial \sigma_{\phi\theta}}{\partial \phi} + \frac{\sigma_{r\theta}}{r} - \frac{\cot\theta}{r}\sigma_{\phi\phi} \\[2mm] \frac{1}{r^2}\frac{\partial}{\partial r}(r^2\sigma_{r\phi}) + \frac{1}{r}\frac{\partial \sigma_{\phi\theta}}{\partial \theta} + \frac{1}{r\sin\theta}\frac{\partial \sigma_{\phi\phi}}{\partial \phi} + \frac{\sigma_{r\phi}}{r} + \frac{2\cot\theta}{r}\sigma_{\theta\phi} \end{pmatrix}$$

Convective derivative of a vector **u**

$$\frac{D\mathbf{u}}{Dt} = \begin{pmatrix} \frac{\partial u_r}{\partial t} + u_r\frac{\partial u_r}{\partial r} + \frac{u_\theta}{r}\frac{\partial u_r}{\partial \theta} - \frac{u_\theta^2+u_\phi^2}{r} + \frac{u_\phi}{r\sin\theta}\frac{\partial u_r}{\partial \phi} \\[2mm] \frac{\partial u_\theta}{\partial t} + u_r\frac{\partial u_\theta}{\partial r} + \frac{u_\theta}{r}\frac{\partial u_\theta}{\partial \theta} + \frac{u_r u_\theta}{r} + \frac{u_\phi}{r\sin\theta}\frac{\partial u_\theta}{\partial \phi} - \frac{u_\phi^2\cot\theta}{r} \\[2mm] \frac{\partial u_\phi}{\partial t} + u_r\frac{\partial u_\phi}{\partial r} + \frac{u_\theta}{r}\frac{\partial u_\phi}{\partial \theta} + \frac{u_\theta u_\phi\cot\theta}{r} + \frac{u_r u_\phi}{r} + \frac{u_\phi}{r\sin\theta}\frac{\partial u_\phi}{\partial \phi} \end{pmatrix}$$

Stokes equations

$$\frac{\partial u_r}{\partial r} + 2\frac{u_r}{r} + \frac{1}{r}\frac{\partial u_\theta}{\partial \theta} + \frac{1}{r\sin\theta}\frac{\partial u_\phi}{\partial \phi} + \cot\theta\frac{u_\theta}{r} = 0$$

$$-\frac{\partial p}{\partial r} + \mu\left[\nabla^2 u_r - \frac{2u_r}{r^2} - \frac{2}{r^2}\frac{\partial u_\theta}{\partial \theta} - \frac{2u_\theta\cot\theta}{r^2} - \frac{2}{r^2\sin\theta}\frac{\partial u_\phi}{\partial \phi}\right] = 0$$

$$-\frac{1}{r}\frac{\partial p}{\partial \theta} + \mu\left[\nabla^2 u_\theta + \frac{2}{r^2}\frac{\partial u_r}{\partial \theta} - \frac{u_\theta}{r^2\sin^2\theta} - \frac{2\cos\theta}{r^2\sin^2\theta}\frac{\partial u_\phi}{\partial \phi}\right] = 0$$

$$-\frac{1}{r\sin\theta}\frac{\partial p}{\partial \phi} + \mu\left[\nabla^2 u_\phi - \frac{u_\phi}{r^2\sin^2\theta} + \frac{2}{r^2\sin\theta}\frac{\partial u_r}{\partial \phi} + \frac{2\cos\theta}{r^2\sin^2\theta}\frac{\partial u_\theta}{\partial \phi}\right] = 0$$

For axisymmetric flows:

- Axisymmetric stream function

$$u_r = \frac{1}{r^2\sin\theta}\frac{\partial\Psi}{\partial\theta}, \qquad u_\theta = -\frac{1}{r\sin\theta}\frac{\partial\Psi}{\partial r}$$

- Axisymmetric Stokes operator $\mathfrak{D}^2$

$$\mathfrak{D}^2\Psi = \frac{\partial^2\Psi}{\partial r^2} + \frac{\sin\theta}{r^2}\frac{\partial}{\partial\theta}\left(\frac{1}{\sin\theta}\frac{\partial\Psi}{\partial\theta}\right)$$

Bibliography

[1] D. Barthès-Biesel and A. Acrivos. Deformation and burst of a liquid drop freely suspended in a linear shear field. *J. Fluid Mech.*, 61:1–21, 1973.

[2] D. Barthès-Biesel and J. M. Rallison. The time-dependent deformation of a capsule freely suspended in a linear shear flow. *J. Fluid Mech.*, 113:251–267, 1981.

[3] D. Barthès-Biesel and H. Sgaier. Role of membrane viscosity in the orientation and deformation of a capsule suspended in shear flow. *J. Fluid Mech.*, 160:119–135, 1985.

[4] G. K. Batchelor. Slender-body theory for particles of arbitrary cross-section in Stokes flow. *J. Fluid Mech.*, 44:419–440, 1970.

[5] G.K. Batchelor. The stress system in a suspension of force-free particles. *J. Fluid Mech.*, 41:545–570, 1970.

[6] G.K. Batchelor. Transport properties of two-phase materials with random structure. *Ann. Rev. Fluid Mech.*, 6:227–255, 1974.

[7] G.K. Batchelor and T. Green. The determination of the bulk stress in a suspension of spherical particles to the order c-square. *J. Fluid Mech.*, 56:401–427, 1972.

[8] N. Baumann, D. D. Joseph, P. Mohr, and Y. Renardy. Vortex rings of one fluid in another in free fall. *Phys. Fluids A*, 4:567, 1992.

[9] G . S. Beavers and D. D. Joseph. The rotating rod viscometer. *J. Fluid Mech.*, 69:475–511, 1975.

[10] T.B. Benjamin. Wave formation in laminar flow down an inclined plane. *J. Fluid Mech.*, 2:254–274, 1957.

[11] S. Bhattacharya and J. Blawzdziewicz. Image system for Stokes-flow singularity between two parallel planar walls. *J. Math. Physics*, 43:5720–5731, 2002.

[12] J.F. Brady and G. Bossis. The rheology of concentrated suspensions of spheres in simple shear flow by numerical simulation. *J. Fluid Mech.*, 155:105–129, 1985.

[13] J.F. Brady and G. Bossis. Stokesian dynamics. *Ann. Rev. Fluid Mech.*, 20:111–157, 1985.

[14] C. Bucherer, C. Lacombe, and J.C. Lelievre. *Biomécanique des Fluides et des Tissus*, pages 31–51. Masson, 1998.

[15] A.T. Chwang and T.Y.T. Wu. Hydrodynamics of low Reynolds number flow. Part 1: Rotation of axisymmetric prolate bodies. *J. Fluid Mech.*, 63:607– 622, 1974.

[16] A.T. Chwang and T.Y.T. Wu. Hydrodynamics of low Reynolds number flow. Part 2: Singularity methods for Stokes flow. *J. Fluid Mech.*, 67:787–815, 1975.

[17] P. Coussot. *Rheometry of Pastes, Suspensions and Granular Materials.* Wiley-Interscience, New York, 2005.

[18] M. Coutanceau. Sur l'étude expérimentale de l'écoulement engendré par une sphère en écoulement de Stokes. *C.R.A.S.*, A274:853, 1972.

[19] W.R. Dean and P.E. Montagnon. On the steady motion of viscous liquid in a corner. *Proc. Camb. Phil. Soc.*, 45:389–394, 1949.

[20] A. Diaz and D. Barthès-Biesel. Entrance of a bioartificial capsule in a pore. *CMES*, 3(3):321–337, 2002.

[21] A. Einstein. Eine neue Bestimmung der Molekldimensionen. *Ann. Phys.*, 19:289– 306, 1906.

[22] D.R. Foss and J.F. Brady. Structure, diffusion and rheology of Brownian suspensions by Stokesian dynamics simulation. *J. Fluid Mech.*, 407:167–200, 2000.

[23] N.A. Frankel and A. Acrivos. On the viscosity of a concentrated suspension of solid spheres. *Chem. Eng. Sci.*, 22:847–853, 1967.

[24] N.A. Frankel and A. Acrivos. The constitutive equation for a dilute emulsion. *J. Fluid Mech.*, 44:65–78, 1970.

[25] J. D. Goddard and C. Miller. Nonlinear effects in the rheology of dilute suspensions. *J. Fluid Mech.*, 28:657–673, 1967.

[26] C. Hancock, E. Lewis, and H.K. Moffatt. Effects of inertia in forced corner flows. *J. Fluid Mech.*, 112:315–327, 1981.

[27] J.J.L. Higdon. Stokes flow in arbitrary two-dimensional domains: Shear flow over ridges and cavities. *J. Fluid Mech.*, 159:195–206, 1985.

[28] E.J. Hinch and L.G. Leal. The effect of Brownian motion on the rheological properties of a suspension of non spherical particles. *J. Fluid Mech.*, 52:638–712, 1972.

[29] H.E. Huppert. The propagation of two-dimensional and axisymmetric viscous gravity currents over a rigid horizontal surface. *J. Fluid Mech.*, 121:43–58, 1982.

[30] J.B. Jeffery. The motion of ellipsoidal particles immersed in a viscous fluid. *Proc. Roy. Soc. A*, 102:161–183, 1922.

[31] M. Kennedy, C. Pozrikidis, and R. Skalak. Motion and deformation of liquid drops, and the rheology of dilute emulsions in shear flow. *Computers Fluids*, 23:251–278, 1994.

[32] S. Kim and S.J. Karrila. *Microhydrodynamics*. Butterworth-Heinemann, Boston, MA, 1991.

[33] C.J. Koh and L.G. Leal. The stability of drop shapes for translation at zero Reynolds number through a quiescent fluid. *Phys. Fluids A*, 1:1309–1313, 1989.

[34] C.J. Koh and L.G. Leal. An experimental investigation on the stability of viscous drops translating through a quiescent fluid. *Phys. Fluids A*, 2:2103–2109, 1990.

[35] E. Lac, D. Barthès-Biesel, N. A. Pelekasis, and J. Tsamopoulos. Spherical capsules in three-dimensional unbounded Stokes flow: Effect of the membrane constitutive law and onset of buckling. *J. Fluid Mech.*, 516:303–334, 2004.

[36] G.L. Leal. *Laminar Flow and Convective Transport Processes.* Butterworth-Heinemann, Boston, MA, 1991.

[37] Y. Lefebvre and D. Barthès-Biesel. Motion of a capsule in a cylindrical tube: Effect of membrane pre-stress. *J. Fluid Mech.*, 589:157–181, 2007.

[38] A. Leyrat-Maurin and D. Barthès-Biesel. Motion of a deformable capsule through a hyperbolic constriction. *J. Fluid Mech.*, 279:135–163, 1994.

[39] X. Z. Li, D. Barthès-Biesel, and A. Helmy. Large deformations and burst of a capsule freely suspended in an elongational flow. *J. Fluid Mech.*, 187:179–196, 1988.

[40] M.J. Martinez and K.S. Udell. Axisymmetric creeping motion of drops through circular tubes. *J. Fluid Mech.*, 210:565–591, 1990.

[41] H.K. Moffatt. Viscous and resistive eddies near a sharp corner. *J. Fluid Mech.*, 18:1–18, 1964.

[42] A. H. Nayfeh. *Perturbation Methods*. Wiley-Interscience, New-York, 2000.

[43] A. Okagawa, R.G. Cox, and S.G. Mason. The kinetics of flowing dispersions. VI transient orientation and rheological phenomena of rods and discs in shear flow. *J. Colloid Interface Sci.*, 45:303–329, 1973.

[44] C. Pozrikidis. *Boundary Integral and Singularity Methods for Linearized Viscous Flow.* Cambridge University Press, Cambridge, UK, 1992.

[45] C. Pozrikidis. Finite deformation of liquid capsules enclosed by elastic membranes in simple shear flow. *J. Fluid Mech.*, 297:123–152, 1995.

[46] C. Pozrikidis. *Introduction to Theoretical and Computational Fluid Dynamics.* Oxford University Press, Oxford, UK, 1997.

[47] I. Proudman and J.R.A. Pearson. Expansions at small Reynolds number of the flow past a sphere and a circular cylinder. *J. Fluid Mech.*, 2:237–262, 1957.

[48] S. Ramanujan and C. Pozrikidis. Deformation of liquid capsules enclosed by elastic membranes in simple shear flow: Large deformations and the effect of capsule viscosity. *J. Fluid Mech.*, 361:117–143, 1998.

[49] H. A. Stone. Dynamics of drop deformation and breakup in viscous fluids. *Ann. Rev. Fluid Mech.*, 26:65–102, 1994.

[50] P. Tabeling. *Introduction to Microfluidics.* Oxford University Press, Oxford, UK, 2005.

[51] S. Taneda. Visualization of separating Stokes flows. *J. Phys. Soc. Japan,* 46:1935–1942, 1979.

[52] R.I. Tanner. *Engineering Rheology.* Oxford University Press, Oxford, UK, 1985.

[53] G.I. Taylor. *Collected Papers of Sir Geoffrey Ingram Taylor (Ed. G.K. Batchelor), 1971,* volume 4, pages 397–404. Cambridge University Press, Cambridge, UK, 1962.

[54] J. Walter, A.-V. Salsac, D. Barthès-Biesel, and P. Le Tallec. Coupling of finite element and boundary integral methods for a capsule in a Stokes flow. *Int. J. Num. Meth. Engng*, 83:829–850, 2010.

[55] Y. Xia and G. Whitesides. Soft lithography. *Ann. Rev. Mater. Sci.,* 28:153–184, 1998.

[56] G. H. Youngren and A. Acrivos. On the shape of a gas bubble in a viscous extensionnal flow. *J. Fluid Mech.*, 76:433–442, 1976.

[57] H. Zhou and C. Pozrikidis. Deformation of capsules with incompressible interfaces in simple shear flow. *J. Fluid Mech.*, 283:175–200, 1995.

[58] H. Zhou and C. Pozrikidis. Adaptative singularity method of Stokes flow past particles. *J. Computational Phys.*, 117:79–89, 2002.

Index